UNDERSTANDING PHASE DIAGRAMS

Other Engineering titles from Macmillan

John: An Introduction to Engineering Materials
Radford and Richardson: Production Engineering Technology, 2nd edition
Radford and Richardson: The Management of Production, 3rd edition
Redford, Rimmer and Titherington: Mechanical Technology, 2nd edition
Redford: Mechanical Engineering Design, 2nd edition. Revised in SI units

Understanding PHASE DIAGRAMS

V. B. JOHN M.Sc., C.Eng., M.I.M.M., A.I.M.

Senior Lecturer in Engineering
The Polytechnic of Central London

Macmillan

First published 1974 *by*
THE MACMILLAN PRESS LTD
London and Basingstoke
Associated companies in New York Dublin
Melbourne Johannesburg and Madras

SBN 333 14989 0

Typeset in Great Britain at
PREFACE LIMITED
Salisbury, Wilts.,
and printed by
ARROWSMITHS
Bristol

Contents

Preface

My colleagues and I have observed that many students, in particular students of engineering, seem to find great difficulty in understanding the principles of phase diagrams. In consequence, I thought that there could be many students who would appreciate the existence of a monograph on this subject. Normally the student of engineering is only informed about the *p–t* diagram for water and *t–c* phase diagrams for binary alloy systems. I believe that phase diagrams will be more readily appreciated if the subject is given some unity and to this end I have chosen to include vapour pressure curves and *t–c* diagrams for liquid mixtures in this small work. While the book is intended primarily for those following courses in engineering or metallurgy, it is to be hoped that students in other disciplines might find the volume of some interest and value.

Chapter 5, which deals with some specific alloy systems, seeks only to indicate the relationships which exist between phase diagrams, structure and properties, including the existence of metastable phases, for a few major systems. It is not intended that this chapter be a comprehensive review as there are numerous full definitive works available which deal with the properties of alloys.

I would like to thank my colleague, Clive Beesley, for his assistance with manuscript checking, and I am greatly indebted to my wife for her patience and understanding during the gestation period of this work and for converting an often almost illegible manuscript into type.

V. B. John

1

Phase Systems

1.1 Introduction

The terms solid, liquid and gaseous (or vapour) are used to describe the various states of matter. For a particular substance it is also common practice to quote a melting or freezing point, and a boiling or condensation point, but it would be an oversimplification to assume that the substance can only exist as a vapour at temperatures above the boiling point. Consider the interface between a liquid and free space. At any instant some molecules are escaping from the surface of the liquid into space. At the same time some of the gaseous molecules are impacting with the liquid surface and are being captured, that is condensing from vapour into liquid. When the rate of escape of liquid molecules is equal to the rate of condensation the liquid will be in equilibrium with its vapour. Two states of matter will be in coexistence. An increase in temperature will increase the total energy of this liquid–gas system. An increase in the kinetic energy of the molecules will cause more molecules to escape from the liquid per unit time. An increase in the number of vapour molecules coupled with increased molecular velocities will give an increase in the pressure of the vapour and a new position of equilibrium will be established.

The above is an example of a physical reaction involving the interrelationship between the various states of a substance.

Some solid substances can exist in more than one form, for example there are two crystalline forms of carbon, namely graphite and diamond. Similarly, iron can exist in two different crystalline forms.

The relationships between the various states of a substance and the effects of temperature and pressure on these states can be shown by means of diagrams, known as phase or equilibrium diagrams.

Phase diagrams are not confined to simple substances (one-component systems). They may also be produced to show the relationships between two or three substances, as for example between the component metals in an alloy system. The ability to understand and interpret such diagrams is necessary for several branches of science and technology. Phase reactions occur in all fields and many such reactions are of considerable technical significance, for example, solid-state transformation in alloy systems causing major property changes, fractional distillation of petroleum and other chemical liquids, phase relationships affecting the structure and properties of furnace slags. Also, the science (or art) of meteorology is concerned to a large extent with atmospheric phase reactions.

1.2 Phases

A *phase* may be defined as a portion of matter which is homogeneous. Mechanical subdivision of a phase will produce small portions indistinguishable from one another. A phase does not have to be a single substance. Gases mix freely with one another in all proportions to give a fully homogeneous mixture. The gaseous state will therefore always be classed as a single phase irrespective of the number of gases present. In the cases of liquids and solids, the number of phases present will depend upon solubilities. Petrol and water do not mix, so that when one is added to the other there will be two liquid phases, the less dense petrol forming a separate layer above the denser water. Alcohol and water, however, dissolve in one another in all proportions and so any mixture of these liquids constitutes a single phase. Similarly, an unsaturated solution of salt in water is homogeneous and is a single phase. But if the amount of salt is increased beyond the saturation limit the system will consist of two phases with a saturated solution of salt in water existing in equilibrium with excess solid salt.

1.3 Components

Phase systems may be classified as one-component, two-component (binary) or three-component (ternary). What is meant by the term *component* and how does a component differ from a phase? The number of components in a phase system is the smallest number of atomic or molecular species needed to specify all the phases of the system. This statement needs to be clarified by examples. The phase system of ice, water and water vapour is a one-component system, the component being water, H_2O. The fact that water is a compound of hydrogen and oxygen does not affect matters as water does not dissociate into its constituents under normal conditions. In the case of the alloy system of the metals copper and zinc there are six different solid phases, all possessing different crystal structures but this is a two-component system as all phases can be expressed in terms of copper and zinc.

1.4 The Phase Rule

The phase rule was enunciated in 1876 by J. W. Gibbs. This may be expressed symbolically as

$$P + F = C + 2$$

where P is the number of phases
C is the number of components in the system
F is the number of degrees of freedom, or *variance.*

The term *degrees of freedom* requires some explanation. The number of degrees of freedom is the number of parameters, temperature, pressure or composition which can be varied independently without altering the number of phases present. The significance of this statement will be considered more fully in chapter 2.

What is the purpose and meaning of the phase rule? All systems possessing the same number of degrees of freedom behave in a similar manner when subjected to changes in the variables, temperature, pressure or composition. The behaviour patterns may, therefore, be predicted for a very wide range of physical systems. The phase rule, however, gives no information on the rate of any phase reaction.

1.5 Thermodynamic Considerations

A system is in a state of equilibrium when there is no net change occurring. In the mechanical sense this could mean a body at rest in its position of lowest potential energy but it also includes systems involving opposing reactions, as with a liquid being in equilibrium with its vapour when the rate of condensation of vapour is equal to the rate of vaporisation of the liquid. When a system is not in a state of equilibrium, changes, either physical or chemical, will occur†.

†A non-equilibrium state may be metastable and show no apparent change if the total energy of the system is very low.

In thermodynamic terms a system is said to be in a state of equilibrium when the free energy F of the system is a minimum. The free energy of a system may be defined as

$$F = E - TS$$

where E is the internal energy of the system
S is the entropy of the system
T is the temperature

Entropy is a measure of the randomness of a system. According to the Second Law of Thermodynamics a spontaneous change will always take place in such a way as to cause an increase in entropy. Equilibrium in a system will be achieved when the free energy F is low and the entropy S is large.

1.6. Metastable States

Very often a system will exist virtually indefinitely in a state which is not the stable state in terms of lowest free energy. For example, in the chemical system hydrogen, oxygen and water, it is water which possesses the lowest free energy at normal temperatures. On this basis the reaction

$$2H_2 + O_2 \rightarrow 2H_2O$$

should take place spontaneously at room temperature. However, the two gases hydrogen and oxygen can be in intimate contact with one another indefinitely at ordinary temperatures with no reaction occurring. Similarly martensite, a constituent formed in quenched steels, is not an equilibrium state, yet this structure may be retained in quenched steels held at room temperature.

In these and many other cases it is necessary to raise the temperature of the system for the reaction to proceed. The quantity of energy which has to be put in to the system before the reaction will take place is termed the *activation energy* of the process. A mechanical analogy is the case of a tetragonal prism of mass m (figure 1.1).

The prism will remain in position A indefinitely if left undisturbed, even though position C is the position of lowest potential energy for the prism. If the prism is moved from position A to position C by pivoting about point O, it will be seen that the centroid of the figure will follow the path GG′G″. When in position B the potential energy of

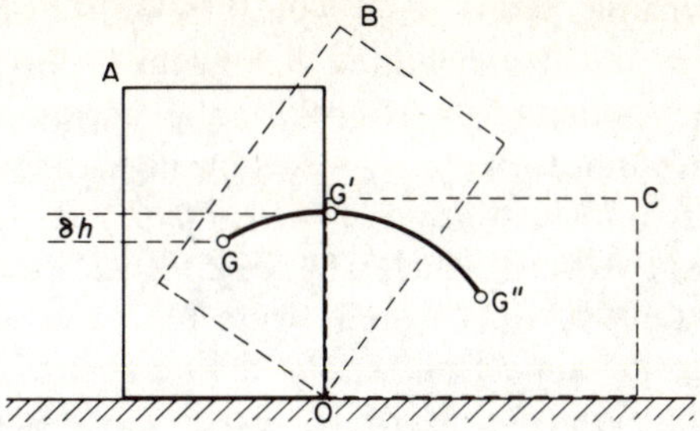

Figure 1.1. *Activation of a process – mechanical analogy*

the prism will be greater than the potential energy of state A by an amount $mg\delta h$. This quantity of energy would be termed the activation energy necessary for the change in prism position to occur.

Within any material the atoms or molecules are continually in motion and possess kinetic energy (atomic motion in solids is vibration about relatively fixed points). The total energy content of a system is determined by the temperature, but not all the atoms or molecules within a system possess the same energy at any particular instant. At any instant in time some molecules may be at rest while others are moving at relatively high velocities. The distribution of energies between the atoms or molecules in a system is given by the Maxwell–Boltzmann distribution law. This is illustrated in figure 1.2. $N(E)$ is the number of atoms per unit volume having an energy within

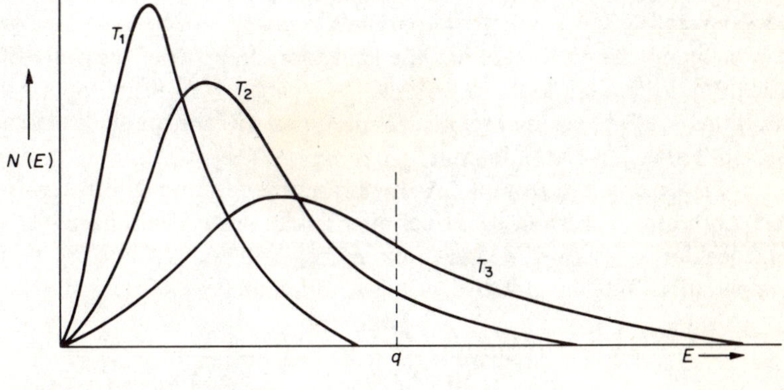

Figure 1.2. *Maxwell–Boltzmann distribution of molecular energies in a system at three temperatures ($T_1 < T_2 < T_3$)*

the range E to $E + \mathrm{d}E$, where $\mathrm{d}E$ is a small increment of energy. Figure 1.2 shows how the N(E) distribution varies with temperature ($T_1 < T_2 < T_3$). As the temperature rises, the total energy of the system increases and the number of atoms possessing high kinetic energy values increases. But from the definition of $N(E)$, $\int_0^\infty N(E)\,\mathrm{d}E = N$, the number of atoms per unit volume, which will remain constant. Therefore, the areas under the three curves shown in figure 1.2 are the same. If q is the activation energy of some process then the number of atoms per unit volume possessing an energy of q, or greater, will be given by $\int_q^\infty N(E)\,\mathrm{d}E$. This function would have a value of zero at temperature T_1, but definite values at temperatures T_2 and T_3. It follows that the reaction would not take place at temperature T_1, but would occur at the temperatures T_2 and T_3. Further, the rate at which the process would occur would be greater at T_3 than T_2 as a greater number of atoms possess energies above level q at this higher temperature.

The rate at which a process occurs is governed by the Arrhenius rate law, which can be written as

$$\text{rate} = \mathrm{A}\exp\left(-\frac{q}{kT}\right)$$

where q is the activation energy for a single molecule, k is Boltzmann's constant, T is the temperature (K) and A is a constant. This may be rewritten as

$$\text{rate} = \mathrm{A}\exp\left(-\frac{Q}{R_\mathrm{o}T}\right)$$

where Q is the activation energy per kilomole, and R_o is the universal gas constant (8.314 kilo joule/kilo mole K).

For many chemical reactions the activation energy is of the order of 40 000 kilo joule/kilo mole. Activation energies for many physical reactions in alloy systems are much higher than this and are in the range of 150 000 to 200 000 kilo joule/kilo mole.

It will be apparent from the Arrhenius rate equation that a change in temperature will exert an enormous effect on the rate of reaction.

Consider a reaction of activation energy 150 000 kJ/kmol at two temperatures 300 K and 900 K.

$$\text{At 900 K} \qquad \exp\left(-\frac{Q}{RT}\right) = \text{-}\exp\left(-\frac{150 \times 10^3}{8.314 \times 900}\right)$$

$$= \mathrm{e}^{-20} \approx 10^{-8.5}$$

$$\text{At 300 K} \qquad \exp\left(-\frac{Q}{RT}\right) = \exp\left(-\frac{150 \times 10^3}{8.314 \times 300}\right)$$

$$= e^{-60} \approx 10^{-25.5}$$

In other words, the reaction would occur approximately 10^{17} times faster at 900 K than at 300 K.

If such a reaction is completed in a time of one second at 900 K, it would require 10^{17} seconds or more than 10^9 years to reach completion at 300 K (room temperature). This is effectively a reaction rate of zero at room temperature. In this way non-equilibrium states may be retained indefinitely provided the temperature is low enough. This is the basis for many metallurgical heat treatments.

The net energy change in a reaction, ΔH, is termed the heat of reaction. The heat of reaction may be an energy absorption, as in endothermic reactions or an energy release, $-\Delta H$, as in exothermic reactions. The relationship between the heat of reaction ΔH and the activation energy Q for endothermic and exothermic reactions is shown in figure 1.3.

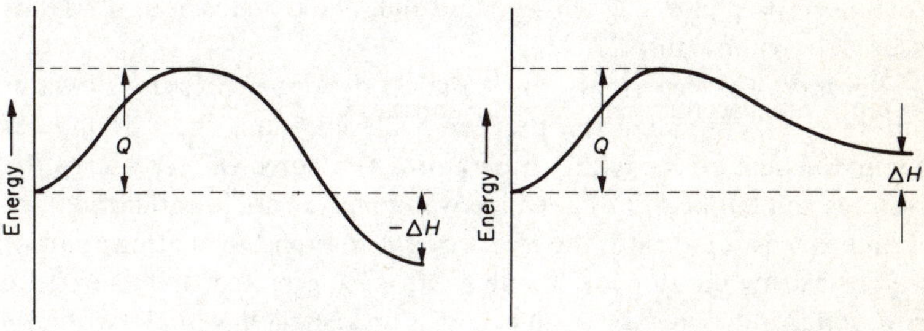

(a) Exothermic reaction
Q = activation energy
$-\Delta H$ = energy emitted

(b) Endothermic reaction
Q = activation energy
ΔH = energy absorbed

Figure 1.3. *Energy change in a reaction*

2

One-Component Systems

2.1. Water

A system consisting of a pure substance, a one-component system, may be represented by a phase or equilibrium diagram with pressure and temperature as the two axes. (It is customary to plot pressure as the ordinate.) The phase diagram for the solid, liquid and vapour phases of water is shown in figure 2.1.

The curve OA represents the variation of vapour pressure of water with temperature. Similarly the curve BO indicates the variation of vapour pressure of ice with temperature. The curve OC represents the pressures and temperatures at which water and ice are in equilibrium, or in other words it indicates the effect of pressure on the melting point of ice. For clarity the slope of the line OC is exaggerated. If the pressure on a solid and liquid in equilibrium is increased the phase with the larger volume will tend to disappear. This is in accordance with Le Chatelier's principle, which may be stated as follows. *If, for a system in equilibrium, one of the factors such as temperature or pressure is changed then the position of equilibrium will shift in an attempt to offset the effect of the change.* Ice is less dense than water and so an increase in pressure will lead to a decrease in melting point. (An increase of one atmosphere ($10^5 N/m^2$) in pressure will reduce the freezing point of water by 0.0075°C. With most substances the liquid phase in less dense than the solid and an increase in pressure would cause an increase in melting point.)

The curve OA shows that an increase in temperature causes the vapour pressure of water to rise. A liquid is said to boil when its vapour

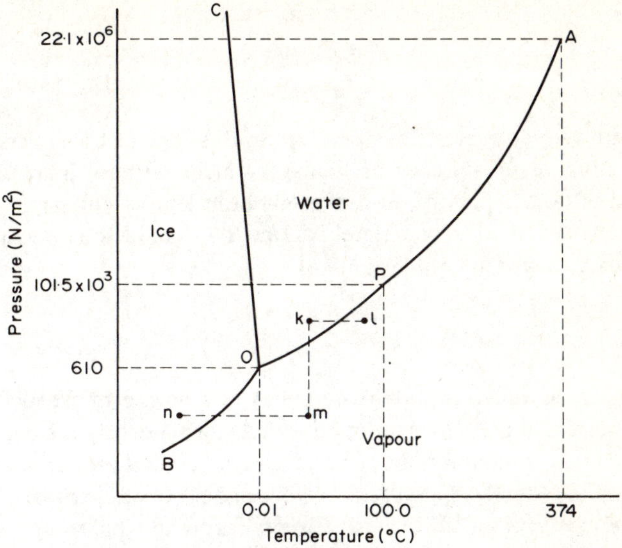

Figure 2.1. *Phase diagram for water*

pressure is equal to the external pressure. Point P on the curve OA represents the normal boiling point of water at 100.0°C (373.13 K) and a pressure of 101.5×10^3 N/m^2 (1 atmosphere). Ice also has a vapour pressure and this, although small, is shown by curve BO. BO is not a continuation of the curve OA, but is a separate curve as it refers to a separate phase. The two vapour pressure curves intersect at O. Point O is a *triple point* since three phases, water, ice and vapour, exist in equilibrium. The vapour pressure of the two phases at the triple point is 610 N/m^2. The normal melting point of a solid and the triple point are not coincident. The normal melting point of a solid is the temperature at which the solid melts at atmospheric pressure. For ice the melting point occurs at 0°C (273.13 K), whereas the triple point, where ice is in equilibrium with water under the pressure of its own vapour, occurs at 0.01°C (273.14 K).

2.2. Degrees of Freedom or Variance

Applying the phase rule to the water system we have the following. For a single phase area, for example the ice area bounded by BOC and the axes

$$P + F = C + 2$$
$$1 + F = 1 + 2$$
$$F = 2$$

There are two degrees of freedom. Within this zone or field, a change in temperature or pressure or both may be made without increasing the number of phases present, or conversely both temperature and pressure have to be specified to completely define the state of the system. When two phases coexist in equilibrium

$$2 + F = 1 + 2$$
$$F = 1$$

The system is univariant; that is, it has one degree of freedom. The temperature and pressure cannot be varied independently if both phases are to remain in equilibrium. Two phases, water and vapour, coexist at points on the line OA in figure 2.1. Only one parameter, temperature or pressure, need be specified when two phases are in equilibrium in order to completely define the system.

Three phases can only coexist in equilibrium at the triple point O.

$$3 + F = 1 + 2$$
$$F = 0$$

The triple point is unique and invariant. The fact that three phases are in equilibrium completely fixes the temperature and pressure of the system.

2.3. The Phase Diagram

Referring to the phase diagram for water, figure 2.1, the diagram consists of areas or fields bounded by the lines OA, OB and OC. Within each field there is only one stable phase and the fields are labelled accordingly, ice, water and vapour. At a boundary line, OA, OB or OC two phases coexist in equilibrium.

As an example of the interpretation of this type of phase diagram consider the state at point k on the diagram with specified values of temperature and pressure. Under these conditions there is only one stable phase, namely water. If the temperature is increased to a value corresponding to point l with no change in pressure the liquid will convert completely to vapour, as vapour is the only stable state at the temperature and pressure specified by point l. Similarly, if from state k the pressure is reduced at constant temperature to a value equivalent to point m there will again be complete vaporisation of the liquid. If from state m the pressure is maintained constant, but the temperature is reduced to a value corresponding to point n there will be direct conversion of vapour to solid without passing through the liquid phase. This is the condition which gives rise to hoar frost deposition, namely a sudden fall in temperature when the pressure of water vapour in the atmosphere is less than 610 N/m^2.

When a transition from one phase to another takes place there will be a change in the internal energy and entropy of the system. If the system is being heated the internal energy and entropy of the new phase formed will be higher than for the old phase stable at lower temperatures. Energy will be absorbed at the transition point and the amount of energy absorbed is termed the latent heat of the transition. Conversely, a phase transition occurring during cooling will be accompanied by the emission of latent heat.

Point A in the diagram is the *critical point*. Beyond this point liquid and vapour phases become identical. Occasionally there are restrictions in applying the phase rule; this is one. The effect of such restrictions is to reduce the number of degrees of freedom by one. Normally for two phases in equilibrium in a one-component system there should be one degree of freedom. In this case a restriction (R) has to be employed because the two phases become identical.

$$P + F = C + 2 - R$$

$$2 + F = 1 + 2 - 1$$

$$F = 0$$

The critical point is invariant and for water has specific values of temperature and pressure (374°C (647 K) and 22.1 x 10^6 N/m^2).

2.4. The Critical Point

The change from vapour to liquid, or from liquid to vapour generally occurs quite suddenly. A series of isothermal curves is shown in figure 2.2. At some low temperature T_1 the behaviour of a gas departs considerably from 'ideal' behaviour as stated by the relationship

$$pV = RT$$

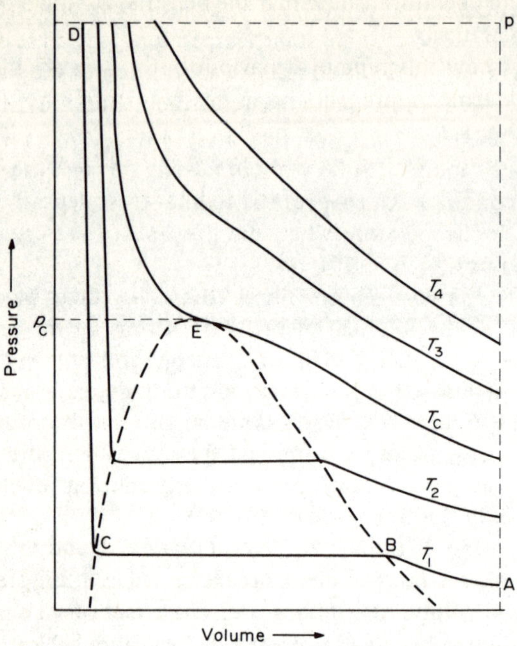

Figure 2.2. *Isothermal curves for a substance* $(T_1 < T_2 < T_c < T_3 < T_4)$

As pressure is increased at temperature T_1 the volume of the gas decreases following the curve AB but at a pressure corresponding to point B the volume suddenly reduces to the low value C as the gas liquefies. A further increase in pressure causes little further reduction in volume as the liquid is not very compressible. This is shown by the portion CD on the T_1 isothermal.

In following the path ABCD at a constant temperature the transition from gaseous to liquid state occurs suddenly. It is possible to proceed from A to D by a different route. If the temperature of the vapour is increased from T_1 to a high temperature keeping the volume constant the pressure will rise appreciably, following the path AP. If the pressure is now maintained at a constant value and the temperature reduced to T_1 the path PD will be followed and the substance will have transformed from vapour to liquid, but in a gradual manner with no sharp discontinuity. This indicates that under these conditions there is no difference between gas and liquid. This is referred to as *the continuity of the liquid and gaseous states.* At the critical point E on the critical temperature isothermal the densities of liquid and saturated vapour are identical.

All gases show this type of behaviour but values of critical pressure and critical temperature vary considerably from one substance to another.

2.5. Sublimation

The phase diagrams for other pure substances are similar to that for water, unless the substances show allotropic modifications. Figure 2.3 shows the phase diagram for carbon dioxide, CO_2, and it will be seen that this differs from that for water in two ways. Firstly, the slope of the solid–liquid phase boundary is in the opposite sense to that for water, as in this case solid carbon dioxide is more dense than the liquid. It will also be noted that the pressure at the triple point O is considerably above atmospheric pressure. The vapour pressure of solid carbon dioxide, 'dry ice', is equal to standard atmospheric pressure at a temperature of -78°C (195 K). At atmospheric pressure solid carbon dioxide sublimes, transforming directly from the solid state to the vapour phase without liquefying. Any substance will sublime rather than melt when heated at atmospheric pressure if the triple point pressure is higher than atmospheric.

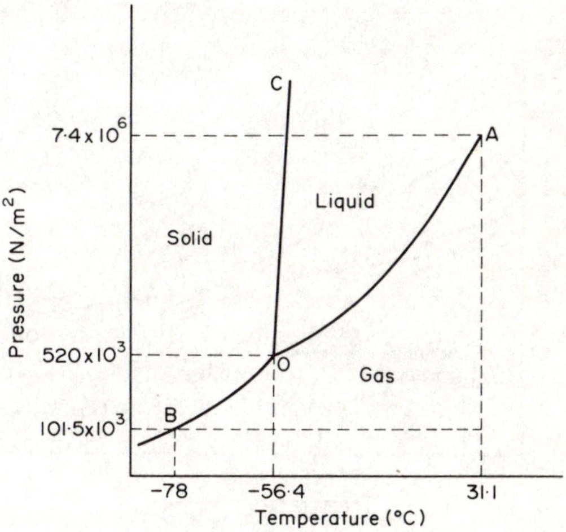

Figure 2.3. *Phase diagram for carbon dioxide*

2.6. Allotropy

Certain solid substances can exist in more than one crystalline form. This is termed *polymorphism* or *allotropy*. Among the elements that exhibit allotropy are carbon, sulphur, tin and iron. Diamond and graphite are two allotropic forms of carbon, while sulphur may crystallise in either the rhombic or the monoclinic form.

The free energy of a substance decreases with an increase in temperature and the greater the specific heat of a substance the more rapid will be the rate of decrease of free energy. For a substance with two possible crystalline forms, α and β, the free energies of the phases α, β, and liquid will vary with temperature in the manner shown in figure 2.4. The α modification will be the stable form at all temperatures up to T_c, since it is the form with the lowest free energy in this temperature range, but at temperatures between T_c and T_m the β modification will be the stable form. Above temperature T_m the liquid will be stable.

This type of polymorphism, with each phase possessing a definite range of stability, is termed *enantiotropy* and it is a relatively common type. The vapour pressure–temperature relationships for such a system are shown in figure 2.5. Curve AB is the vapour pressure curve for the α modification, BC is the vapour pressure curve for the β modification and CD the vapour pressure curve for the liquid. There are

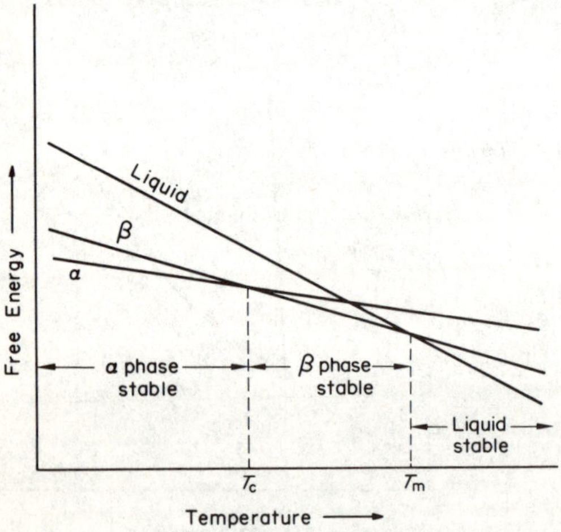

Figure 2.4. *Variation of free energy with temperature for an enantiotropic substance*

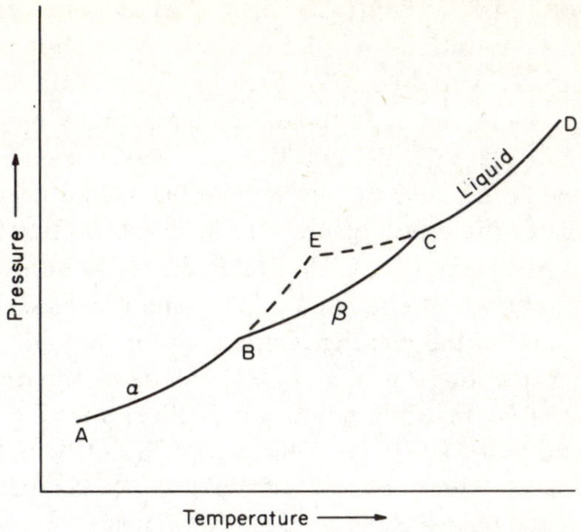

Figure 2.5. *Vapour pressure curve for an enantiotropic substance*

discontinuities at B and C. These are triple points. Slow heating of the α form, allowing equilibrium to be established, will result in the α form transforming to the β form at a temperature corresponding to B. Further rise of temperature will see the vapour pressure of β increasing according to the curve BC with melting occurring at a temperature corresponding to point C. These changes take place in reverse order during cooling.

If the substance is heated rapidly so that equilibrium is not attained, the vapour pressure of the α form will increase beyond B along the extension curve BE. The α form will melt at a temperature corresponding to E. The vapour pressure of the liquid will then continue to rise with further increase in temperature following the curve EC which is an extension of the curve CD. Similarly, rapid cooling will allow liquid to transform directly into the α form at a temperature below the freezing point at which liquid should change into β. Point E is a metastable triple point. The α modification at temperatures above B and liquid at temperatures below C are both metastable.

Enantiotropic behaviour is typified by the sulphur system.

2.7. Sulphur

The stable form of sulphur at ordinary temperatures has a rhombic crystal structure. When rhombic sulphur is heated rapidly it melts at a temperature of 114.5°C (Point E in figure 2.6) but when heated slowly it undergoes a transition at 95.5°C into β sulphur possessing a monoclinic crystal structure (Point B in figure 2.6). Further slow heating will cause melting of β sulphur at a temperature of 119.25°C.

In figure 2.6 lines AB, BC and CD are vapour pressure curves for stable states while BE and EC are metastable vapour pressure curves. Line BF indicates the effect of pressure on the transition temperature for the α to β transition. Line CF indicates the effect of pressure on the melting temperature of monoclinic sulphur and line EFG indicates the effect of pressure on the melting temperature of rhombic sulphur, the portion EF representing the metastable melting of the rhombic form. Points B, C and F are triple points where three phases exist in stable equilibrium and point E is a metastable triple point. It is theoretically impossible for all four phases of sulphur to coexist in stable equilibrium. If the phase rule is applied it would give

$$4 + F = 1 + 2$$

$$F = -1$$

A variance of −1 is clearly an impossible situation and so rhombic sulphur, monoclinic sulphur, liquid and vapour cannot all coexist in stable equilibrium.

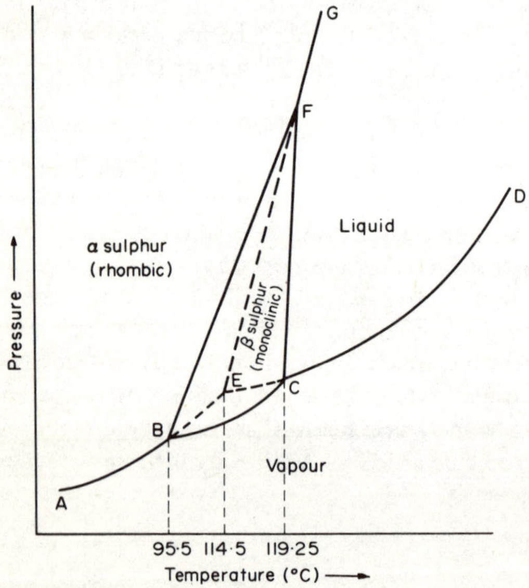

Figure 2.6. *Phase diagram for sulphur*

2.8. Monotropy

Some substances exhibit a type of polymorphism known as *monotropy*. A monotropic substance is one that possesses more than one crystalline form but where one form is stable over the whole temperature range and the other form is merely metastable at all temperatures. In figure 2.7a curve AB is the vapour pressure curve of the α modification. B is the triple point very close to the normal melting point and BC is the vapour pressure curve of liquid. Curve FED is the vapour pressure curve of the β form. Extrapolation of the curve AB will give an intersection with the vapour pressure curve of β at D. Point D may be thought of as being the transition temperature, T_c, for the transformation from α to β, but this point is purely hypothetical as it lies well above the melting points of both crystalline forms. The vapour pressure curve for the β form lies above that for the α form indicating that the β form is always metastable (see also the free-energy curves figure 2.7b.)

The β form cannot be created by direct transformation from α as this would be against all the laws of thermodynamics. If, however, the temperature of the liquid is reduced rapidly the liquid will be retained in a metastable condition at temperatures below B, the vapour pressure of liquid following the path BE. At a temperature corresponding to point E the liquid will solidify into the β form. β may also be produced by rapid cooling of vapour at a low pressure giving a 'hoar frost' type deposition. The β modification, being metastable, will always have the tendency to transform into α, but in many cases the rate of transformation at ordinary temperatures may be so slow that, to all

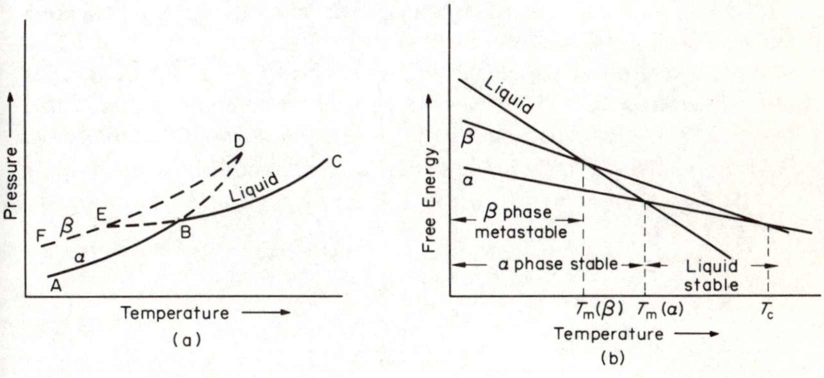

Figure 2.7. *(a) Vapour pressure curves for a monotropic substance. (b) Free energy curves for a monotropic substance*

intents and purposes, the β form will remain unchanged. Phosphorus and carbon are monotropic. White phosphorus and diamond are the metastable forms of these two elements. At ordinary temperatures white phosphorus will slowly change to red phosphorus. In diamond on the other hand, the rate of change from metastable to the stable form graphite is infinitesimal and no change occurs at normal temperatures.

Some substances are both monotropic and enantiotropic. A good example is silica, SiO_2. α and β quartz are two stable forms of silica showing enantiotropic behaviour. α and β cristobalite are two metastable forms of silica which are enantiotropic with respect to one another but which bear a monotropic relationship to the two stable forms, α and β quartz.

2.9. Iron

A number of metals show allotropic modifications. Of these, the most commercially important is iron. Iron can exist in two crystalline forms, body-centred cubic and face-centred cubic. The crystal structure of iron is body-centred cubic at all temperatures from zero up to 908°C (1181 K). This form is termed α iron. On heating beyond 908°C the structure of iron changes to face-centred cubic, a more closely packed and hence denser state. This form is termed γ iron which remains the stable form up to 1388°C (1661 K) when the structure reverts to the body-centred cubic form. The high-temperature body-centred cubic structure is termed δ iron but it is crystallographically identical with α. The δ iron is stable at temperatures up to the melting point of 1535°C (1808 K). α iron loses its ferromagnetic characteristics on heating above 768°C (1041 K) and early workers used the term β to describe the state of iron at temperatures between 768°C and 908°C. When it was discovered that there was no crystallographic change associated with the loss of magnetism, use of the term β iron was discontinued. As will be seen later the presence of allotropic modifications in a pure substance will have an effect on the form of phase diagrams for binary systems.

3

Binary Liquid Systems

3.1. Representation of a Binary System

With a system involving two components it is necessary that composition be indicated on a phase diagram. The complete diagram would be based on three orthogonal axes, pressure, temperature and composition, respectively and would therefore be a three-dimensional space diagram. Such a *p–t–c* diagram is difficult to depict and use is made of separate plane diagrams, pressure–temperature, pressure–composition and temperature–composition. In the latter two cases the base line, or abscissa, of the diagram is used to indicate composition changes with pressure or temperature as the ordinate. In a *p–c* or *t–c* diagram the base line shows all possible compositions of the two components, A and B, from 100 per cent of one component to 100 per cent of the other. Composition is usually represented as percentage by weight (w/o), but it is equally valid to used the concept of molar fraction as the basis for defining the composition of any particular mixture (figure 3.1). When considering a binary phase diagram between two elements, as in metallic alloy systems, the base line may represent either composition by percentage weight or composition by atomic percentage (a/o) the latter being analagous to the use of molar fractions.

3.2. Liquid Mixtures with Complete Solubility

When two liquids are brought together they may:

(a) dissolve completely in one another in all proportions;
(b) partially dissolve in one another;
(c) be completely immiscible.

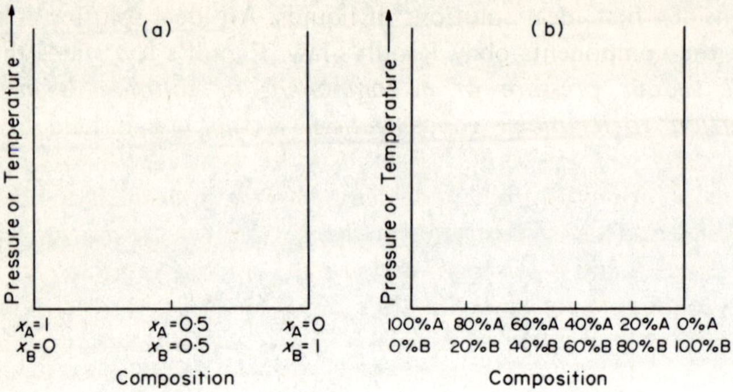

Figure 3.1. *Representation of composition in a phase diagram: (a) by molar fraction; (b) by percentage (w/o or a/o)*

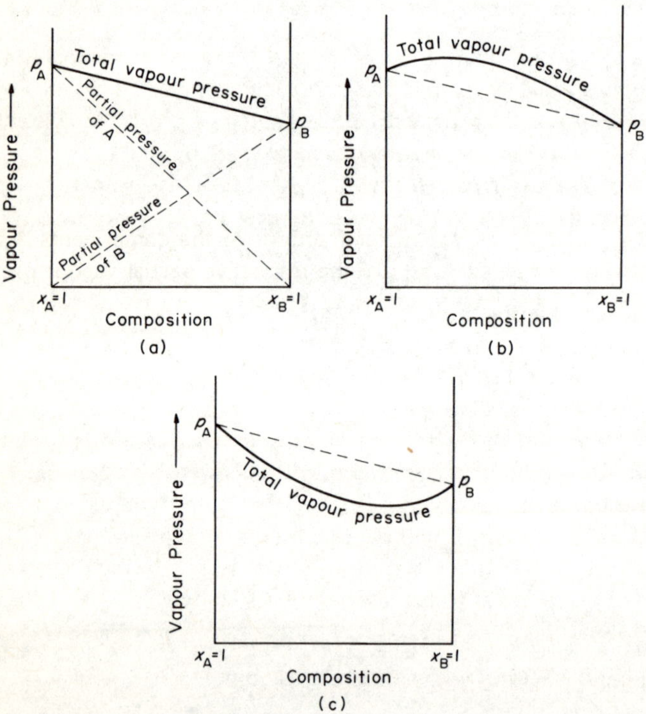

Figure 3.2. *Variation of vapour pressure with composition for a binary mixture: (a) ideal mixture obeying Raoult's law; (b) non-ideal mixture showing positive deviation; (c) non-ideal mixture showing negative deviation*

Consider first ideal solutions of liquids. An ideal solution is one in which the components obey Raoult's law. Raoult's law states that *the partial vapour pressure of a component in solution is in direct proportion to its molar concentration.* If this law is held by both components in a mixture it means that the total vapour pressure of a series of solutions in a binary system will vary in a linear manner with composition. This is illustrated in figure 3.2a for the mixture of two ideal liquids A and B of vapour pressures p_A and p_B respectively.

In the majority of cases Raoult's law is only obeyed when solutions are dilute and mixtures of ordinary liquids show deviations from the law. Such deviations may be positive or negative.

In a non-ideal mixture, that is ordinary liquids, the composition of vapour mixture is not generally the same as the composition of the liquid mixture from which it is derived. Duhem and Margules, in the latter part of the nineteenth century, derived expressions to show a connection between the partial vapour pressures of the components in a binary solution and their concentrations. Their work may be summarised in the equation

$$\frac{x_A}{p_A} \times \frac{dp_A}{dx_A} = \frac{x_B}{p_B} \times \frac{dp_B}{dx_B} \qquad (3.1)$$

where x_A and x_B are the molar fractions of the components A and B respectively and p_A and p_B are the respective partial vapour pressures of A and B. For a binary system of A and B, dx_A, where dx_A is a small incremental change in the concentration of A, must be equal to $-dx_B$

$$\therefore \qquad \frac{x_A}{p_A}\frac{dp_A}{dx_A} = -\frac{x_B}{p_B}\frac{dp_B}{dx_A} \qquad (3.2)$$

If the total vapour pressure of the solution is P, then the slope of the total vapour pressure curve is

$$\frac{dP}{dx_A} = \frac{dp_A}{dx_A} + \frac{dp_B}{dx_A}$$

$$= \frac{dp_B}{dx_A}\left(1 + \frac{dp_A}{dp_B}\right) \qquad (3.3)$$

From (3.2)

$$\frac{dp_A}{dp_B} = -\frac{x_B p_A}{x_A p_B}$$

Substituting in (3.3)

$$\frac{dP}{dx_A} = \frac{dp_B}{dx_A}\left(1 - \frac{x_B p_A}{x_A p_B}\right) \tag{3.4}$$

The value of dp_B/dx_A is equal to $-dp_B/dx_B$ and is therefore negative. Hence if dP/dx_A is to be positive, from equation (3.4) $x_B p_A$ must be greater than $x_A p_B$: that is

$$\frac{p_A}{p_B} > \frac{x_A}{x_B}$$

This means that the vapour mixture contains a higher concentration of A than the liquid mixture with which it is in equilibrium. Conversely, if dP/dx_A is negative the vapour will be richer in component B than is the liquid solution.

Only if the total vapour pressure curve for the binary system shows a maximum or minimum point will the vapour mixture and the solution contain the same proportions of the components A and B. For a maximum or minimum point

$$\frac{dP}{dx_A} = 0$$

Therefore

$$\frac{x_B p_A}{x_A p_B} = 1$$

or

$$\frac{p_A}{p_B} = \frac{x_A}{x_B}$$

That is, the composition of vapour mixture is the same as the composition of the liquid mixture with which it is in equilibrium when there is a maximum or minimum on the vapour pressure–composition curve.

As stated earlier, the normal boiling point of a liquid is the temperature at which the total vapour pressure of the liquid is equal to 101.5×10^3 N/m^2 (atmospheric pressure). This statement applies equally to a liquid mixture as to a pure substance. For an ideal mixture the variation of boiling point with composition would be linear as the total vapour pressure curve is a straight line. For non-ideal mixtures, the boiling point curve will not be linear but will be the inverse of the total vapour pressure, as a high vapour pressure indicates a low boiling point and vice versa. The matter is further complicated because, as shown above, in the general case the composition of vapour mixture will not be the same as the composition of the liquid mixture in equilibrium with it.

Phase diagrams may be plotted from experimentally observed data to show the relationship for liquid mixtures. In figure 3.3 are shown a pressure–composition diagram and a temperature–composition diagram for a typical non-ideal liquid mixture, in which the vapour pressure–composition curve shows neither a maximum nor a minimum. In the example illustrated in figure 3.3 the slope of the vapour pressure curve dP/dx_A is positive and so, for a liquid–vapour system in equilibrium, the vapour phase is richer in component A than the liquid phase.

Consider the boiling of a liquid mixture (see figure 3.3b). When the temperature of a liquid mixture of composition X is raised to temperature t_1, that is point m on the lower curve AB, the liquid mixture will begin to boil. The composition of the vapour-phase mixture that is evolved will correspond to point p on the upper curve AB. As the vapour phase formed is richer in component A than the original liquid mixture the composition of the remaining liquid will be enriched in the component B and consequently the boiling point of the remaining mixture will increase. When the temperature is raised to t_2 liquid of composition n will boil evolving a vapour of composition q. If

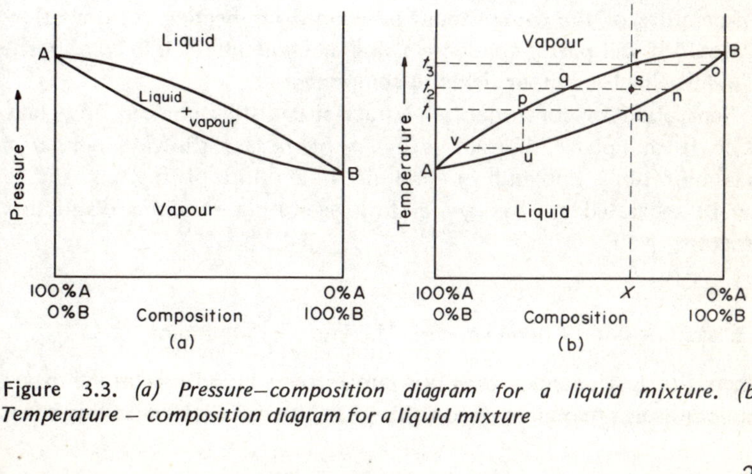

Figure 3.3. *(a) Pressure–composition diagram for a liquid mixture. (b) Temperature – composition diagram for a liquid mixture*

the vapour is kept within a closed system at constant pressure allowing for equilibrium between liquid and vapour to be attained at each successive temperature, the last of the liquid, of composition o, will vapourise at temperature t_3 converting the composition of the total vapour mixture to r. In figure 3.3b the lower curve AB, the boiling point curve, is termed the *liquidus* and the upper curve AB is termed the *vaporus.*

During the distillation of a liquid mixture the conditions described above do not occur as there is not a closed system and vapour is continuously removed from the vicinity of the boiling liquid and condensed. Consider the distillation of a liquid mixture of composition X. When boiling commences vapour of composition p is evolved. If this first vapour fraction is led away and condensed it will condense to a liquid of composition u. Redistillation of this fraction will produce a first vapour fraction of composition v, much richer in component A than the original. At the same time residual liquid in the distillation chamber is becoming richer in component B. This has great practical significance. Repeated fractional distillation of a liquid mixture, with this type of boiling point curve will give almost complete separation of the constituents in the mixture. The efficiency of the separation process can be increased by use of a fractionating column. In a fractionating column the vapour rising from the boiling liquid is partially condensed at the various levels in the column. The less volatile (that is, higher boiling point) fraction tends to condense leaving the more volatile vapour to rise up the column and escape over into the collector. Within the column there is an upward stream of vapour meeting a downward flowing liquid at a lower temperature. When the temperature of the rising vapour is reduced on meeting the downflow of liquid it will partly condense giving a condensate rich in component B and leaving the vapour richer in component A.

Separation of constituents by fractional distillation can only take place if the phase diagram is as in figure 3.3 showing neither a maximum nor a minimum. Liquid air is a mixture of this type and so can be separated into oxygen and nitrogen by a fractional distillation process.

3.3. Boiling Point Curves showing Maximum or Minimum

When the total vapour pressure curve for a liquid mixture shows a minimum or a maximum (see figure 3.2) the variation of boiling point with composition curves will show a maximum or a minimum point. It has already been shown that when $dP/dx_A = 0$ the compositions of liquid and vapour phases will be identical. Therefore, at a maximum or minimum point the liquidus and vaporus curves will be coincident.

Phase diagrams for these types of system are shown in figure 3.4. In this figure the lower curves in each case are the liquidus curves while the upper curves are the vaporus. The major difference between these diagrams and figure 3.3b is that the liquid composition denoted by *M* will boil at a constant temperature and distil without change in composition. Such mixtures are termed *azeotropic* mixtures.

Unlike the mixtures dealt with in the preceding section, the constituents of an azeotropic mixture cannot be separated by fractional distillation.

It must be remembered that these temperature–composition phase diagrams for liquid mixtures are based on constant pressure data (normally atmospheric pressure) and a change in external pressure will cause a change in the boiling point of any particular mixture. In the case of azeotropic mixtures (points M in figure 3.4a and b) a change in the external pressure will cause a change in both the boiling point and the composition of the mixture.

3.4. Interpretation of Phase Diagrams

Applying the phase rule to the type of binary diagram discussed in the above sections we have the following

(a) For a single-phase region

$$P + F = C + 2$$

$$1 + F = 2 + 2$$

$$F = 3$$

Three parameters, temperature, pressure and composition must be specified to fully define the state of the system.

(b) For a two phase region

$$2 + F = 2 + 2$$

$$F = 2$$

In this case if two parameters, say temperature and pressure are specified, the compositions of the two phases are also defined. Or one can say that if one parameter, say pressure, is maintained constant then

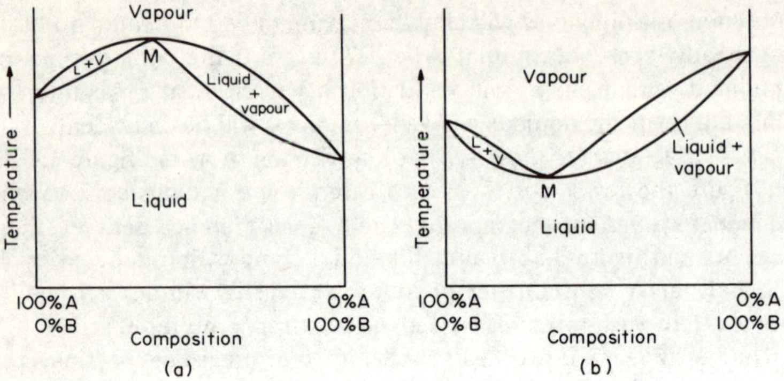

Figure 3.4. *(a) Phase diagram showing maximum boiling point. (b) Phase diagram showing minimum boiling point*

any change in another parameter, temperature, will cause a change in phase composition.

(c) For an azeotropic mixture (point M is figure 3.4) a restriction must be introduced as both phases are identical in composition.

$$P + F = C + 2 - R$$

$$2 + F = 2 + 2 - 1$$

$$F = 1$$

By fixing one parameter only, say pressure, the state will be fully defined. Composition and boiling point will remain constant unless the pressure is altered.

It is appropriate at this stage to introduce some other simple rules for the interpretation of binary phase diagrams.

(i) A binary phase diagram consists of a series of lines which divide it into a number of areas or fields. These fields may be single phase or two phases. Three phases may only coexist at a unique point.

(ii) Single phase areas are separated by a two phase zone, the only exception to this being when both phases are of the same composition as at point M in figure 3.4.

(iii) When a vertical line representing the composition of some mixture in the system cuts a line in the phase diagram it is an indication that some change is taking place. For example, the X composition line in figure 3.3b cuts phase boundary lines and points m and r indicates that during heating boiling of the mixture begins at temperature t_1 and is completed at temperature t_3.

(iv) For any point within a two-phase region the compositions of the two phases in equilibrium with one another is determined by the intersections of a horizontal tie line with the phase boundary lines. In the temperature–composition diagram figure 3.3b the tie line qn intersects the phase boundary lines at *q* and *n* indicating that at temperature t_2 vapour phase of composition *q* is in equilibrium with liquid phase of composition *n*.

(v) For a two-phase state in equilibrium the relative proportions of phases present can be determined using the lever rule. Again referring to figure 3.3b consider the state *s* specified by composition *X*, temperature t_2 and a specific pressure. Liquid and vapour phases will be present in equilibrium, the quantities of phases being in proportion to the lengths of the lever lines, that is

$$\frac{\text{quantity of liquid (composition } n)}{\text{quantity of gas (composition } q)} = \frac{\text{length } qs}{\text{length } sn}$$

3.5. Liquid Mixtures showing no Miscibility

When two liquids are completely immiscible in one another each will exert its own vapour pressure, and this will be unaffected by the presence of the other liquid. As the vapour pressure of a liquid is independent of the mass of liquid present, the total vapour pressure of an immiscible liquid mixture will be constant at a constant temperature, irrespective of the relative amounts of the two components. As the normal boiling point of a liquid is the temperature at which the total vapour pressure is equal to atmospheric it follows that any mixture of insoluble liquids will boil at a temperature below the boiling point of either component.

This principle is used in the practice of steam distillation in which a high boiling point liquid immiscible, or nearly so, in water may be distilled at a comparatively low temperature. This is a useful technique as some high boiling point liquids suffer some decomposition at temperatures near their normal boiling points. Distillation is carried out by passing steam through the mixture rather than be simply heating an immiscible mixture since the steam bubbling through keeps the mixture thoroughly agitated.

The composition of the vapour phase in equilibrium with an immiscible liquid mixture is constant at constant temperature, irrespective of the relative amounts of each liquid present, and may be easily calculated. The number of molecules of each component in the vapour is in proportion to the vapour pressure of the component. This statement may be written as

$$\frac{n_A}{n_B} = \frac{p_A}{p_B}$$

where n_A and n_B are the number of molecules of A and B and p_A and p_B are the respective partial pressures. The mass of substance present is related to the number of molecules according to the expression

$$m = nM$$

Where m is the mass of substance and M is the molecular mass number.

Therefore

$$\frac{m_A}{m_B} = \frac{n_A M_A}{n_B M_B} = \frac{p_A M_A}{p_B M_B}$$

This means that in a steam distillation process the relative masses of the components distilled over are in direct proportion to both the partial pressures and the molecular mass numbers. This is very useful as the organic liquids involved normally have large molecular weights in relation to water.

Consider as an example the steam distillation of chlorobenzene, C_6H_5Cl. The normal boiling point of chlorobenzene is 132°C but it will steam distil at 91°C. At 91°C the vapour pressure of chlorobenzene is 28.75×10^3 N/m^2 and the vapour pressure of water is 72.4×10^3 N/m^2. The composition of distillate may be calculated as

$$\frac{\text{mass of chlorobenzene } (m_A)}{\text{mass of water } (m_B)} = \frac{p_A M_A}{p_B M_B}$$

At 91°C

$$p_A = 28.75 \times 10^3 \text{ N/m}^2$$

$$p_B = 72.4 \times 10^3 \text{ N/m}^2$$

Molecular mass number of chlorobenzene (C_6H_5Cl) = 112.6

Molecular mass number of water (H_2O) = 18

$$\frac{m_A}{m_B} = \frac{28.75 \times 10^3 \times 112.6}{72.4 \times 10^3 \times 18} = 2.48$$

This means that just over 71 per cent of the distillate, by mass, will be chlorobenzene even though the vapour pressure of chlorobenzene at 91°C is only 28 per cent of atmospheric pressure.

3.6. Liquid Mixtures with Partial Miscibility

In between the two extremes already discussed, namely complete solubility of liquids and total immiscibility, there are numerous liquid–liquid systems which show partial miscibility. As an example, if a small amount of phenol is added to water at ordinary temperatures it will completely dissolve. Similarly, if a small amount of water is added to phenol at ordinary temperatures a homogeneous solution of water in phenol will be formed. If, however, water and phenol are mixed together in approximately equal proportions at ordinary temperatures two saturated solutions of differing densities will appear, one an aqueous solution saturated with phenol and the other phenol fully saturated with water. Two solutions of this type are termed conjugate solutions.

In the case of the water–phenol system an increase in temperature will cause an increase in the limits of solubility for both phenol in water and water in phenol. The two solubility curves slope toward one another and merge into each other at a temperature of 68°C at atmospheric pressure (figure 3.5). The point C, at which the two solubility curves merge is termed the *consolute temperature* or *critical solution temperature*. In this case, as there is a maximum on the solubility curve, point *C* is known as the upper consolute temperature.

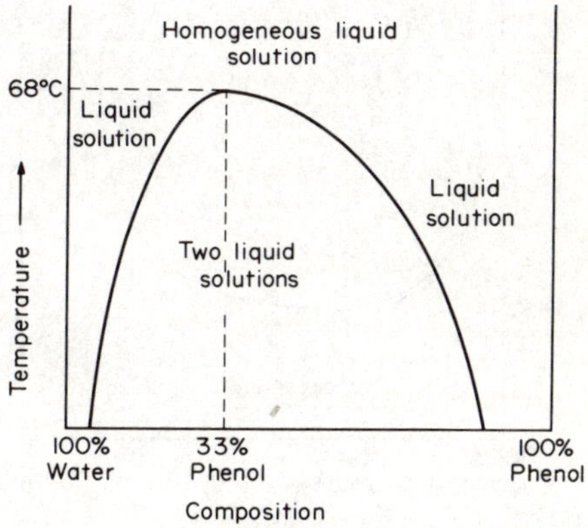

Figure 3.5. *t–c phase diagram for phenol and water at atmospheric pressure*

There are some cases known where the solubility of one liquid in another suffers a decrease as the temperature is raised so giving a lower consolute, or critical solution, temperature. In systems of this type an increase in temperature does not result in an indefinite decrease in liquid solubility and it is found that after a certain point solubility limits tend to increase with an increase in temperature. On this basis a liquid mixture with a lower consolute temperature should also show an upper consolute temperature. This is true in some cases and a good example is the nicotine–water system (figure 3.6). An upper consolute temperature does not occur in all systems because the total vapour pressure reaches atmospheric pressure and hence the liquids boil before the two solubility curves merge. The types of t–c phase diagram which are obtained in these circumstances are summarised in figure 3.7.

Consider the effect of heating a liquid mixture containing X per cent of B. According to figure 3.7a this mixture will consist of two separate liquid solutions at low temperature, the composition of each conjugate solution being given by points F and G respectively. As the temperature is raised the compositions of the two solutions will alter, following the paths FC and GD, as the solubility limits of B in A and A in B increase. Lines FC and GD may be termed solvus lines. At temperature t_1 the sum of the partial pressures of liquid solution 1, of composition C, and

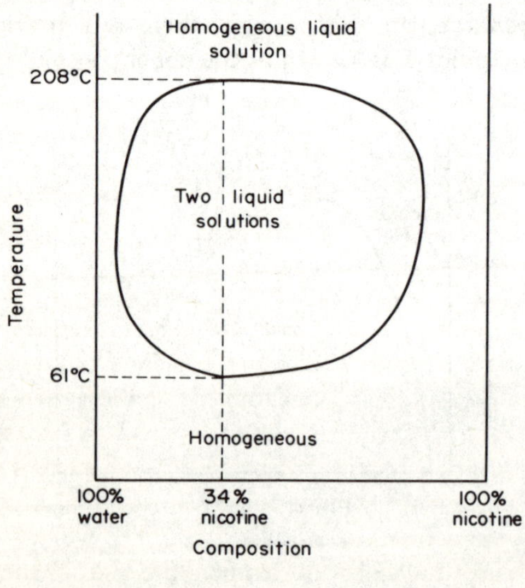

Figure 3.6. *t–c phase diagram for nicotine and water at atmospheric pressure*

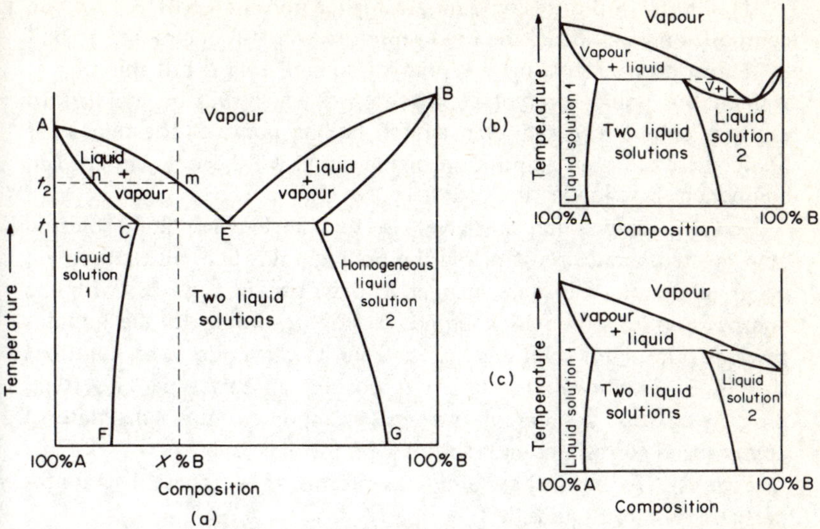

Figure 3.7. *(a) t–c diagram for partially miscible liquids. (b) and (c) alternative diagram shapes*

liquid solution 2, of composition *D*, is equal to atmospheric pressure and the liquid mixture will commence to boil. The line CED forms part of the liquidus line ACEDB. The composition of the vapour mixture evolved as the mixture of liquid solutions boils corresponds to point E on the diagram. Applying the phase rule to point E, there are three phases in equilibrium, vapour and two separate liquids, so

$$P + F = C + 2$$

$$3 + F = 2 + 2$$

$$F = 1$$

Point E is univariant, but if one parameter, pressure, is fixed the temperature and composition of point E will remain constant. Because the percentage of B in the vapour formed is greater than *X* per cent the liquid layer richer in B will be fully vaporised first leaving some of the liquid of composition *C* remaining. Once the system is reduced to a single liquid solution the boiling point will rise, following the liquidus curve CA and the composition of vapour formed will vary according to the vapourus curve EA since the temperature rises in the same manner as described for miscible liquids (page 23).

If a liquid mixture containing a higher percentage of B than the composition at point E, the liquid mixture will still commence to boil at temperature t_1 creating a vapour of composition *E* but this time it will be the liquid layer of composition *C* which will be the first to vaporise completely and then as the boiling point of the remaining liquid increases the composition of the liquid will become enriched in component B following the liquidus curve DB.

For any composition that gives rise to a single-phase liquid solution at some temperature just below the boiling point line, distillation will occur in exactly the same manner as described in page 23, with the composition of the boiling liquid varying according to the liquidus towards pure A for solutions rich in A and towards pure B for solutions rich in B. Fractional distillation is possible in this type of system, progressing to a residue of one pure component but a distillate of composition corresponding to point E on the diagram.

Figures 3.7b and c show alternative forms of *t–c* phase diagram for partially miscible liquids.

4

Liquid-Solid Systems

4.1. Condensed Systems

When considering the transition from liquid to solid for binary systems the effects of pressure can generally be ignored. It has already been stated that while changes in pressure do affect the melting point of a substance the effect is of a very small order. Consequently, the *t–c* type of diagram is the only type of phase diagram normally considered for liquid–solid systems. Because the effect of pressure is negligible we may consider the system as a condensed system and use a reduced version of the Phase Rule. The reduced Phase Rule may be written

$$P + F = C + 1$$

When two liquids are mixed together, as we have seen already, they may either by completely miscible in one another, be partially soluble in one another, or be completely immiscible. Similarly, when liquids solidify there are several possibilities. The two components of homogeneous liquid solution may be:

(a) totally insoluble in one another when solid;
(b) totally miscible with one another forming a continuous series of solid solutions;
(c) partially soluble in one another when solid;
(d) combine with each other to form one or more compounds.

Temperature–composition phase diagrams are of particular importance in the study of many alloy systems, since most alloys are made in the liquid phase and it is convenient to consider the formation of alloy structures on the basis of the solidification of liquids.

4.2. Total Solid Insolubility

Consider the case of two pure substances, A and B, which are completely soluble in one another in the liquid state, but are totally insoluble in one another in the solid state. If a composition base line and temperature scale is drawn (figure 4.1) certain information can be plotted. The melting point of pure substance A can be marked off as point A on the left-hand temperature axis. Similarly point B on the right-hand axis represents the melting point of pure substance B. At high temperatures any mixture of the two liquids will be a single-phase liquid solution.

In the same way as the presence of dissolved salt depresses the freezing point of water, so the freezing point of a liquid will normally be depressed if the liquid contains some other substance in solution. Line AL in figure 4.1 indicates the depression of freezing point of pure A containing dissolved B. Similarly, line BM is the depression of freezing point curve for pure B containing dissolved A. It is important to note that at any point on line AL it is pure substance A which is freezing, that is during cooling the solid which is forming is crystals of

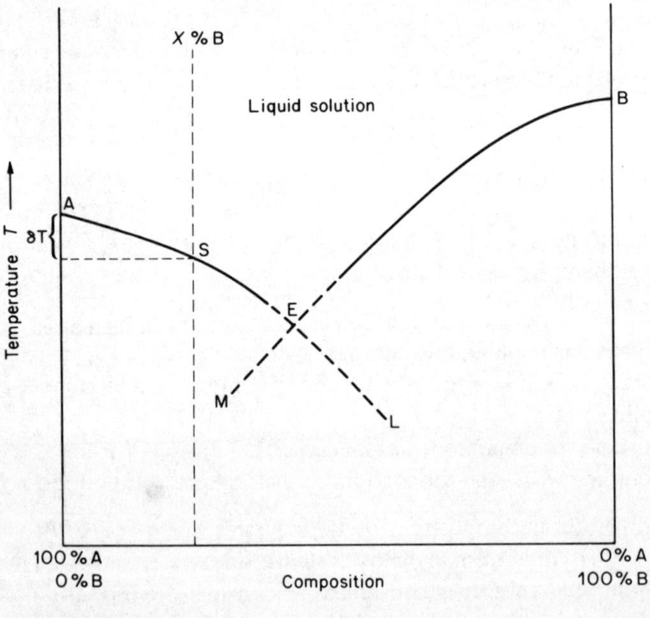

Figure 4.1. *Freezing point curves for two substances insoluble in the solid state*

pure A. For example, X per cent of B dissolved in liquid A will depress the freezing point of A by an amount δT. The liquid solution will begin to solidify at a temperature of A $- \delta T$ (point S on curve AL in figure 4.1) but it is crystals of pure A which will begin to solidify. The two curves AL and BM intersect at point E. The sections of curves EL and EM are hypothetical since no liquid can exist at a temperature lower than that of point E.

Figure 4.2 shows the completed phase, or equilibrium, diagram for the binary system of A and B. Consider again a mixture containing X per cent of B. At a high temperature this exists as a single-phase liquid solution. On cooling the liquid will commence freezing at a point denoted by S on curve AE. Crystals of solid pure A will begin to form. If pure A is rejected from the solution the composition of the remaining liquid must become enriched in B; that is, the composition of the liquid varies toward the right. This means that as the freezing of A continues the temperature and composition of the liquid remaining follows the curve AE toward point E. Point E, which is the only point common to both freezing point curves, represents the lowest temperature that a liquid solution can exist at, and at this point all remaining

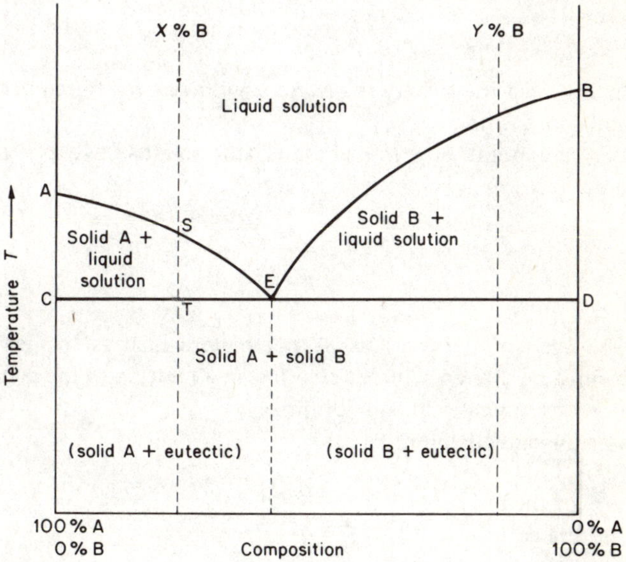

(a)

Figure 4.2. *Binary phase diagram for solid insolubility. (Simple eutectic)*

liquid solution solidifies forming a fine crystal-grained mixture of both solids A and B. Point E is termed the *eutectic point* and the fine-grained crystal mixture formed is termed the *eutectic mixture.* The final structure of the solid mixture containing X per cent of B will, therefore, be composed of large crystals of pure A (primary crystals) and a eutectic mixture of A and B.

If a liquid solution containing Y per cent of B is allowed to solidify, solidification would follow a similar pattern, but in this case primary crystals of pure B would solidify first. It is important to note that the composition of the eutectic mixture remains constant.

In the phase diagram line AEB is termed the *liquidus* and line CED is termed the *solidus*. At all points above the liquidus the mixture is always liquid, and below the solidus the mixture is always wholly solid. Between liquidus and solidus, in the solidification range, the mixture is in a pasty stage.

Applying the reduced phase rule to this type of system we have:

(a) for any point above the liquidus a single-phase liquid

$$P + F = C + 1$$
$$1 + F = 2 + 1$$
$$F = 2$$

In this area the system is bivariant with respect to temperature and composition.

(b) For any point between liquidus and solidus, two phases are present so

$$2 + F = 2 + 1$$
$$F = 1$$

The system is univariant. A parameter, such as temperature, cannot be altered without creating an alteration in the composition of the phases in equilibrium.

(c) For the eutectic point E

$$3 + F = 2 + 1$$
$$F = 0$$

The system is invariant and the eutectic point is unique, with fixed values for temperature and composition.

There are a number of binary systems that form a simple eutectic mixture as shown above. These include metallic alloy systems and systems involving organic compounds.

Many metals solidify from liquid in a dendritic manner. Solidification commences at a nucleus and outward growth from the nucleus occurs preferentially in three directions. Subsequently, secondary and then tertiary arms grow producing a skeleton-type crystal, as in figure 4.3. Outward growth ceases when the advancing dendrite arms meet an adjacent crystal. When outward growth has ceased the dendrite arms thicken and eventually the whole mass is solid and no trace of the dendritic formation remains, except where shrinkage causes inter-dendritic porosity, or in alloy systems where the final liquid to solidify is of a different composition from the primary dendrites.

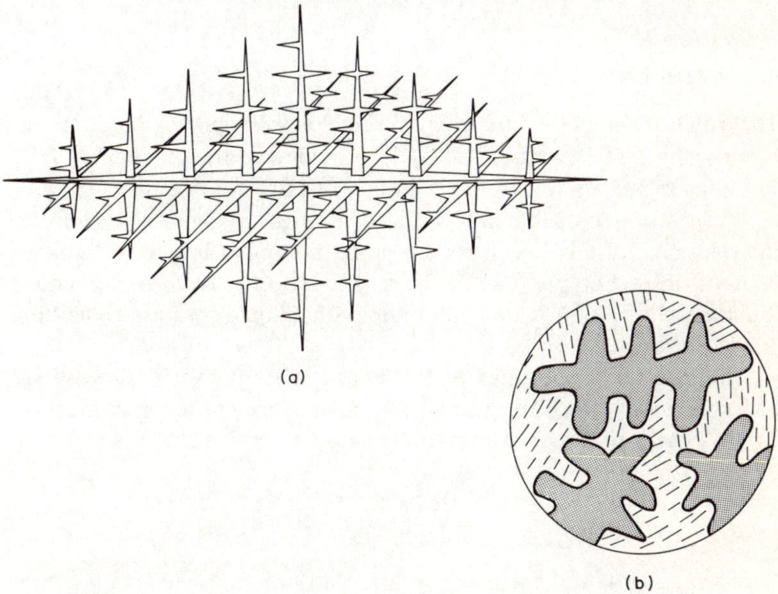

Figure 4.3. *(a) Representation of a dendrite. (b) Solid structure of a simple eutectic alloy. Dendrites of A in eutectic mixture*

An extreme example of the simple eutectic is the case in which the liquidus is a continuous line from the melting point of component A to that of component B. This system, which is not common, is termed the monotectic. Figure 4.4 shows the monotectic phase diagram for the silicon–tin system.

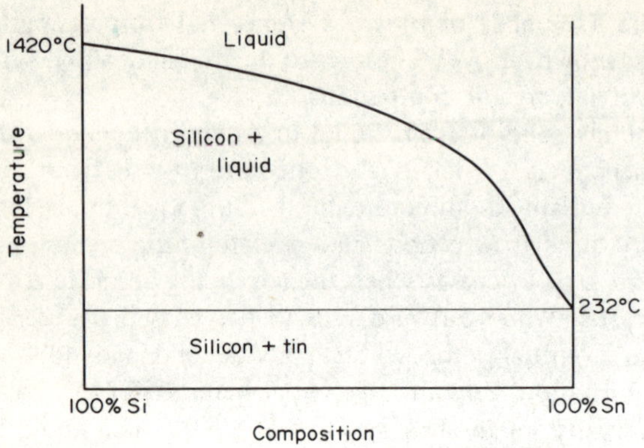

Figure 4.4. *Phase diagram for silicon–tin (monotectic)*

4.3. Interpretation

The rules for the interpretation of liquid–solid phase diagrams are exactly the same as those for liquid–liquid systems (see section 3.4). For a further example of interpretation refer to figure 4.5.

When a horizontal tie line is drawn through a two-phase region the intersections of this line with the phase boundary lines denote phase compositions. For point U in figure 4.5 the intersections at x and y indicate that solid A is in equilibrium with a liquid solution containing *y* per cent of B.

The relative proportions of the phases present can be determined using the lever rule. The quantities of phases present are in proportion to the lengths of the lever lines, for example for point U in figure 4.5

$$\frac{\text{quantity of solid A}}{\text{quantity of liquid (composition } y)} = \frac{\text{Uy}}{\text{Ux}}$$

Similarly, at point V the phases present are solids A and B in the ratio

$$\frac{\text{quantity of A}}{\text{quantity of B}} = \frac{\text{Vr}}{\text{Vp}}$$

Alternatively it could be considered that the phases present are solid A plus eutectic mixture in the ratio

$$\frac{\text{quantity of A}}{\text{quantity of eutectic}} = \frac{\text{Vq}}{\text{Vp}}$$

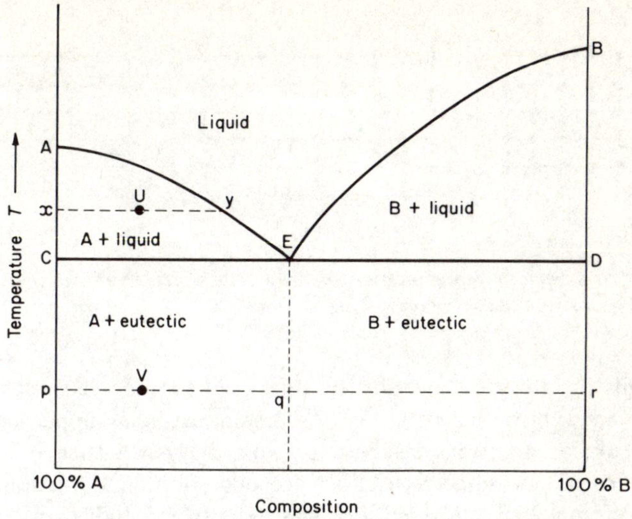

Figure 4.5. *Application of lever rule*

or the percentage of eutectic mixture in the solid mixture of composition V is given by

$$\frac{Vp}{pq} \times 100$$

4.4. Solid Solubility

It is possible for solids to form what is termed a *solid solution*. This may apply whether the solids involved are elements or compounds. For simplicity the following text refers to solutions of metallic elements but the general principles of a common crystal lattice in a single-phase solid solution applies equally to compounds.

The concept of a solid solution may seem strange to some readers, but it simply means that the atoms of the two elements have taken up positions in a common crystal lattice forming a single phase. The atoms of one element enter into the space lattice of the other element in either an interstitial or substitutional manner, as in figure 4.6. The arrangement of dissolved atoms is normally random, but in some instances substitutional solid solutions of an ordered type may be formed. An ordered solution (also known as a superlattice) can only exist at one fixed composition.

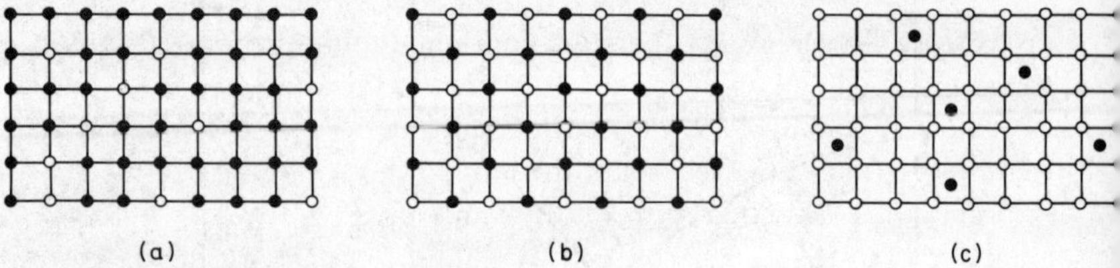

Figure 4.6. *Schematic representation of solid solutions: (a) substitutional (random); (b) substitutional (ordered); (c) interstitial*

Atoms in interstitial or substitutional solid solution cause strain to be developed in the parent lattice. As there must be an upper limit to the amount of strain that can be tolerated in a crystal lattice, it follows that there will be some restrictions to solid solution formation. The nature of metallic solid solutions was extensively studied by Hume-Rothery, and his work is summarised in the following 'rules'.

(a) *Relative size*. If the sizes of the atoms of two metals do not differ by more than 14 per cent, conditions are favourable for the formation of substitutional solid solutions. If the relative sizes of atoms differ by more than 14 per cent solid-solution formation, if it occurs at all, will be extremely limited. Interstitial solid solutions may be formed if the atoms of the solute element are very small in comparison with those of the solvent metal.

(b) *Chemical affinity*. When two metals have a high affinity for one another the tendency is for solid solubility to be severely restricted and intermetallic compounds to be formed instead. This occurs when one element is electronegative and the other is electropositive.

(c) *Relative valency*. If a metal of one valency is added to a metal of another valency the number of valency electrons per atom, the electron ratio, will be altered. Crystal structures are very sensitive to a decrease in the electron ratio. Consequently, a metal of high valency can dissolve very little of a metal of low valency, although a metal of low valency might be able to dissolve an appreciable amount of a high-valency metal.

(d) *Crystal type*. If two metals are of the same crystal-lattice type and all other factors are favourable it is possible for complete solid solubility to occur over the whole composition range. (It is also necessary that the relative sizes of atoms differ by not more than 7 per cent for complete solid solubility.)

4.5. Phase Diagram for Total Solid Solubility

For a binary system where there is a continuous range of solid solution formed the possible phase diagram shapes are as shown in figure 4.7.

A solution containing X per cent of B, figure 4.7a, would solidify in the following manner. Freezing of the liquid solution would commence at temperature t_1. At this temperature liquid of composition l would be in equilibrium with a solid solution of a composition corresponding to point p on the solidus, so the first solid solution crystals to form are of composition p. Consequently, the composition of the remaining liquid becomes enriched in B and the freezing temperature falls slightly. As the temperature falls so the composition of the solid solution tends to change by a diffusion process following the solidus line toward B. At some temperature t_2 liquid of composition m is in equilibrium with

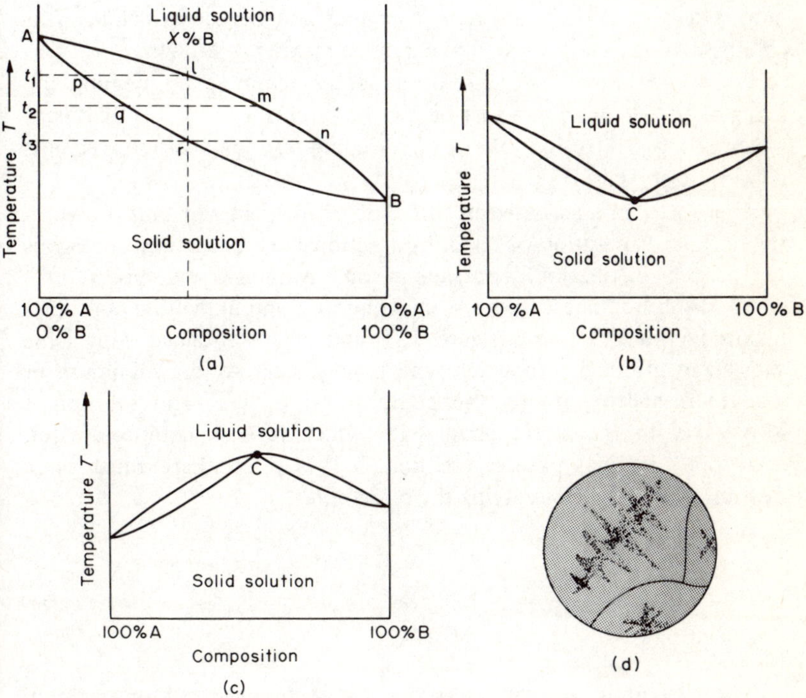

Figure 4.7. *(a) Phase diagram for complete solid solubility. (b) and (c) Alternative phase diagrams for this type. (d) Cored crystal structure*

solid solution of composition q. Solidification will be complete at temperature t_3 when the last drops of liquid, of composition n, solidify, correcting the composition of solid solution crystals to r. If the solidification rate is very slow, allowing for the attainment of equilibrium at all stages during the cooling process, the final solid-solution crystals will be uniform in composition. Normally, however, solidification rates are too rapid for full equilibrium to be attained and the crystals will be cored. In a cored crystal the composition is not the same at all points. The crystal lattice is continuous but there will be a gradual change in composition across the crystal. The centre of the crystal will be rich in substance A while the outer edges will be rich in B. In some metallic alloy systems the coring of crystals is clearly visible under microscopical examination. With alloys of copper and nickel, for example, where the alloy colour is dependent on composition, the centres of crystals are rich in nickel and silvery in appearance while the outer edges of crystals are rich in copper and darker in colour. This colour shading clearly shows the dendritic manner of growth.

Coring in alloys may be subsequently removed by heating the material to a temperature just below the solidus. During this treatment – annealing – diffusion takes place evening out composition gradients within the crystals.

There is a parallel between this type of diagram and that shown in figure 3.3b. Separation of solids by fractional crystallisation, analogous to fractional distillation, is possible in some systems of this type.

Solid-solution phase diagrams showing a minimum melting point or a maximum melting point (figure 4.7b and c) are obtained with some substances although the latter type is very rare. At the minimum or maximum melting points the composition of the solid solution is identical with the liquid phase with which it is in equilibrium (cf. azeotropic liquid mixtures, section 3.3) and so there must be a restriction term when applying the phase rule

$$P + F = C + 1 - R$$

$$2 + F = 2 + 1 - 1$$

$$F = 0$$

The system is invariant and the minimum or maximum point is unique. Although a minimum point in this type of diagram is, like a eutectic, invariant such a minimum melting point is not a eutectic as the solid phase is a homogeneous solution rather than a mixture.

The phase diagram for copper and gold is of the type shown in figure 4.7b. The melting point of copper is 1083°C, that of gold is 1063°C and the minimum melting point alloy contains 81.5 per cent of gold and melts at 884°C.

4.6. Partial Solid Solubility

It is far more common to find that solids are partially soluble in one another rather than be either totally insoluble or totally soluble. A phase diagram for a binary system showing partial solid solubility is given in figure 4.8. This diagram is, in effect, a combination of the two previous types and shows solid solubility sections and also a eutectic. The liquidus is line AEB and the solidus is ACEDB. Lines FC and GD are *solvus* lines and denote the maximum solubility limits of B in A and of A in B respectively. As there are two separate solid solutions formed the Greek letters α and β are used to identify them.

Consider the solidification of three compositions in this system. For mixture (1) solidification begins at temperature t_1 with the formation of β solid solution of composition *O*. As cooling continues the composition of the liquid varies along the liquidus toward point E and the composition of the solid β varies according to the solidus toward

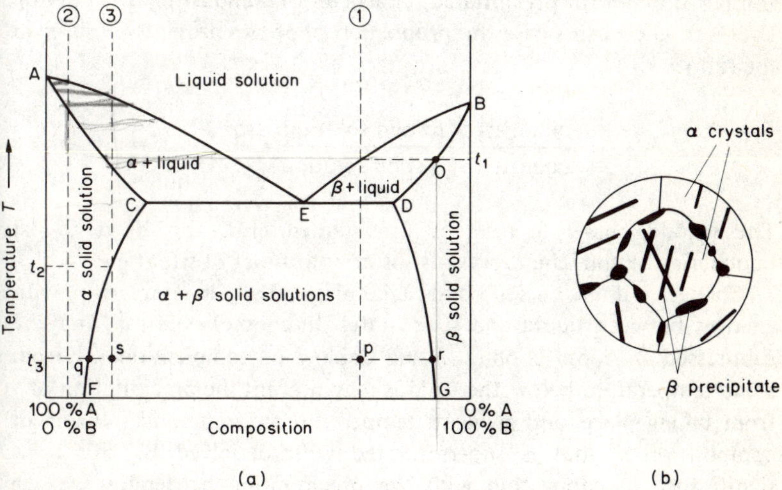

Figure 4.8. *(a) Phase diagram for partial solid solubility with eutectic; (b) structure of composition (3)*

point D. When the eutectic temperature is reached there will be primary cored crystals of β and liquid of the eutectic composition. This liquid then freezes to form a eutectic mixture of two saturated solid solutions, α of composition C and β of composition D. During further cooling the compositions of the α and β phases will adjust, following the solvus lines, until eventually at point p saturated α crystals of composition q will be in equilibrium with saturated β solid solution of composition r.

For mixture (2) solidification of the liquid solution takes place in the same manner as for a complete solid solution (section 4.5) and when solidification is complete the structure will be one of cored α crystals.

In the case of mixture (3) a new concept emerges, namely the possibility of structural changes occurring within the solid state. The liquid solution will freeze on cooling to give a cored α solid solution. During further cooling below the solidus the α solid solution will remain unchanged until temperature t_2 is reached. At this temperature the composition line meets the solvus and the solid solution is fully saturated with component B. As the temperature falls below t_2 the solubility limit is exceeded and excess component B is rejected from solution in A as a precipitate. In this case it is not pure B which forms as a second solid phase, but rather, saturated β solid solution. Eventually, at temperature t_3 the structure is composed of α crystals of composition q with precipitated β particles of composition r. Applying the lever rule in this case the proportion of phases present would be in the ratio

$$\frac{\text{quantity of } \beta \text{ solid solution}}{\text{quantity of } \alpha \text{ solid solution}} = \frac{\text{sq}}{\text{sr}}$$

The second phase, β, may be precipitated either at the α crystal boundaries, within the α crystals, or at both types of site (figure 4.8b).

Changes within the solid state take place slowly in comparison with changes between liquid and solid states. In consequence they may be suppressed by rapid cooling. Rapid cooling of composition (3) from some temperature below the solidus may prevent the precipitation of β from taking place, and giving at temperature t_3 an α solid solution of composition s, that is supersaturated with dissolved B. This is of significance in connection with the precipitation hardening and age hardening of metallic alloys and will be discussed further in chapter 5.

4.7. Peritectic Diagram

Another form of phase diagram which can occur for systems showing partial solid solubility is the *peritectic* type shown in figure 4.9. The liquidus and solidus lines are AEB and ACDB respectively, and FC and GD are solvus lines. The horizontal line CDE is termed the peritectic line and point D the peritectic point

Consider the cooling of liquid of composition (1). Solidification will commence at temperature t_1 with α solid solution of composition q forming. As freezing continues the composition of the liquid follows the liquidus toward point E and the composition of the solid solution follows the solidus toward point C. When the peritectic temperature is reached liquid of composition E exists in equilibrium with α solid solution of composition C. At this temperature the two phases react together to form β solid solution according to the reaction

$$\alpha(\text{composition } C) + \text{liquid (composition } E) \underset{\text{heating}}{\overset{\text{cooling}}{\rightleftharpoons}} \beta(\text{composition } D)$$

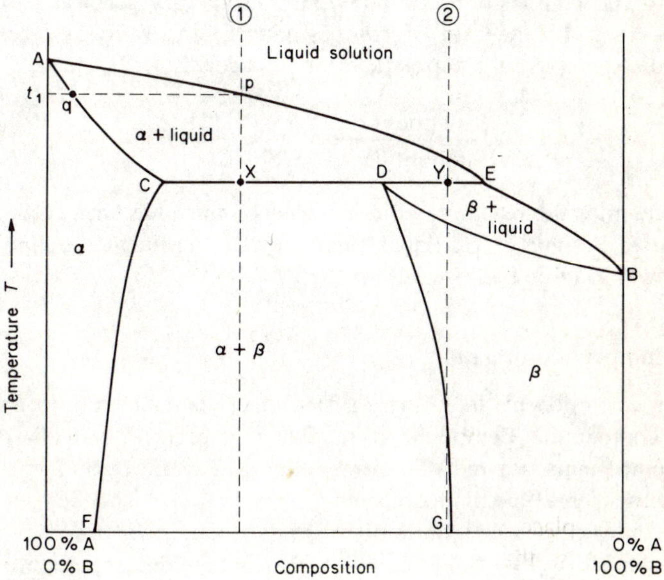

Figure 4.9. *Phase diagram for partial solid solubility with peritectic*

If the reactants α and liquid were present in equivalent proportions, that is in the ratio

$$\frac{\text{amount } \alpha}{\text{amount liquid}} = \frac{DE}{CD},$$

they would both be totally consumed in the reaction producing β solid solution. In the case of composition (1) the reactants were present in the ratio

$$\frac{\text{amount } \alpha}{\text{amount liquid}} = \frac{XE}{CX} \quad \text{(by lever rule)}$$

where

$$\frac{XE}{CX} > \frac{DE}{CD}$$

so that the peritectic reaction will cease when all the liquid is consumed and there is some unreacted α remaining. The structure of the mixture below the peritectic temperature is, therefore, α and β. During further cooling the compositions of both phases will vary according to the solvus lines. In the case of composition (2) the ratio of reactants immediately before the peritectic reaction occurs is

$$\frac{\text{amount } \alpha}{\text{amount liquid}} = \frac{YE}{CY} < \frac{DE}{CD}$$

Consequently the reaction will cease when all the α has been consumed and there is some excess liquid remaining. During further cooling this liquid will solidify as β.

4.8. Compound Formation

The two components in a binary system may combine to form one or more compounds. Compound formation may occur in systems where the components are metallic elements or where the components are themselves organic compounds or inorganic salts. The compounds formed may possess a definite or *congruent* melting point or they may possess an *incongruent* melting point, that is they decompose into one of the components and liquid. This latter is also termed the *meritectic* type.

Compounds are separate phases and possess different crystal

structures from those of the constituents. Generally, intermetallic compounds with congruent melting points possess higher melting points than their constituent metals and they are often brittle and of high hardness. An example is the compound between magnesium and tin, Mg_2Sn.

	Melting Point	Crystal Structure
Magnesium	649°C	close-packed hexagonal
Tin	232°C	body-centred tetragonal
Mg_2Sn	783°C	complex cubic

From many points of view a compound with a congruent melting point can be regarded as a pure substance. The binary phase diagram for a system in which a compound is formed is effectively two simple diagrams linked together. Figure 4.10a shows a phase diagram for two substances A and B, which form one compound, A_xB_y; it is assumed that there is no solid solubility. It comprises two simple binary eutectic diagrams, one between A and the compound and the other between B and compound. Figure 4.10b shows a variation on the above in which there is some solid solubility. As there are now three separate solid solution zones three Greek letters, α, β, and γ, have to be used. Conventionally, these are used progressively working from left to right. When, as in this case, a compound may exist over a small range of composition it is termed an intermediate phase.

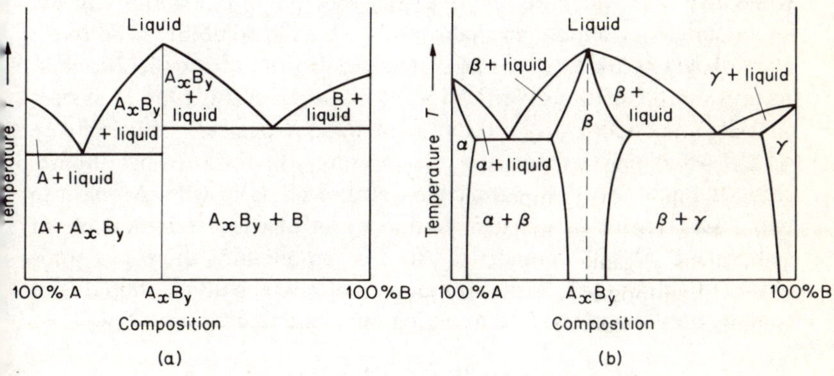

Figure 4.10. *Binary diagrams showing a compound: (a) with no solid solubility; (b) with partial solid solubility*

In some binary systems several compounds or intermediate phases may be formed. Although the complete phase diagram may look highly complex at first sight, it can usually be split into the small and comparatively simple elements discussed in earlier paragraphs, and it can be interpreted according to the same simple rules.

The phase diagrams for systems containing a compound with an incongruent melting point are shown in figure 4.11.

In figure 4.11a the liquidus is AEFB, the solidus is ACEDGHA and point G is the meritectic point, or the incongruent melting point, of the compound A_xB_y. FJ is a hypothetical continuation of curve EF and point J the hypothetical congruent melting point of the compound. Consider the method of solidification of a liquid solution with the same composition as the compound A_xB_y. Solidification will commence when the liquidus is crossed with the freezing of substance B. When the meritectic is reached a meritectic reaction will occur between liquid and B. (Compare peritectics, section 4.7.)

$$\text{Liquid} + \text{B} \underset{\text{heating}}{\overset{\text{cooling}}{\rightleftarrows}} \text{Compound } A_xB_y$$

Figure 4.11b shows an alternative diagram with partial solid solubility. In the magnesium–nickel system, the compound Mg_2Ni has an incongruent melting point.

4.9 Solidification of Partially Miscible and Immiscible Liquid Mixtures

When two substances are only partially miscible in one another in the liquid state they almost invariably show no solid solubility. The likely phase diagrams for this type of system are shown in figure 4.12a and b. As an example of the solidification of mixtures in this type of system consider the cooling of a mixture containing X per cent of B in figure 4.12a. At a high temperature the mixture will form a single homogeneous liquid solution but as the mixture cools it will separate into two liquid layers. When temperature t_1 is reached solidification of component A will commence. At this temperature there are three phases in equilibrium, these are liquid 1 of composition C, liquid 2 of composition D and solid A. Applying the phase rule to point C we have

$$3 + F = 2 + 1$$
$$F = 0$$

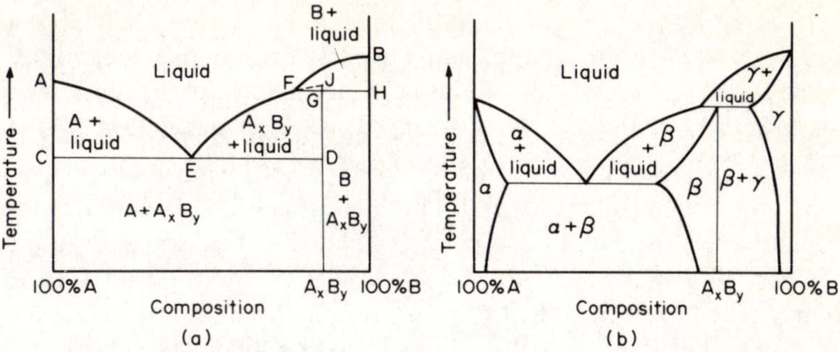

Figure 4.11. *Phase diagrams showing meritectic: (a) with no solid solubility; (b) with partial solid solubility*

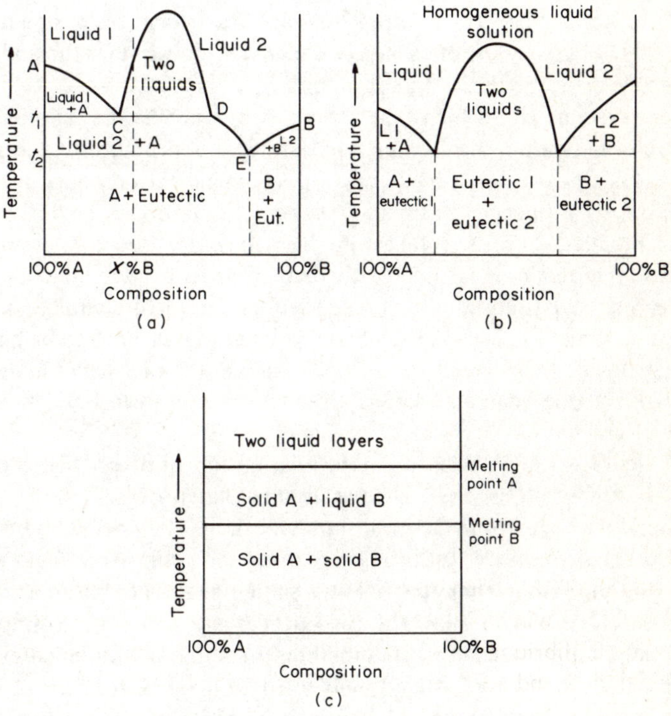

Figure 4.12. *(a) Partially miscible liquids forming one eutectic. (b) Partially miscible liquids forming two eutectics. (c) Phase diagram for total immiscibility.*

Point C is invariant. At temperature t_1 as the solidification of A continues, the relative amount of liquid 2 of composition D increases. The amount of liquid phase 1 steadily diminishes until eventually this liquid layer disappears. Solid A is now in equilibrium with liquid 2 of composition D and the temperature can fall. As the temperature reduces to t_2, the composition of liquid 2 becomes enriched in B following the liquidus curve DE. Point E is a eutectic point and when temperature t_2 is reached all remaining liquid solidifies into the eutectic mixture of components A and B. When two substances are totally immiscible in one another in the liquid state the solids will also be totally insoluble in one another and the phase diagram will be as shown in figure 4.12c.

4.10. Effects of Phase Diagram Type on the Properties of Alloys

Metals are alloyed with one another to produce materials with improved properties. The manner in which the properties of one metal are affected by an addition is largely dependent on whether the additive is miscible or not with the parent metal.

For systems in which the component metals are completely insoluble in one another in the solid state, the structure of the solid alloy is simply a mixture of two pure metals. Consequently, the variation of properties with alloy composition should be linear. In actual practice there is a departure from linearity due to a grain-size effect. A eutectic is a finely divided mixture of two metals. The primary crystals that solidify first on either side of the eutectic point are much larger in size. A fine crystal-grained metal tends to be harder and stronger than a coarse-grained sample of the same material. Similarly, a fine grain size causes a reduction in electrical and thermal conductivities.

In figure 4.13a the approximate relationship between two properties, hardness (H) and electrical conductivity (G), and alloy composition is shown for a simple eutectic alloy. The dotted lines show the expected property variation, neglecting the grain-size effect. Yield strengths and tensile strengths follow a similar pattern to hardness.

In a solid solution alloy the presence of the solute atoms imposes strain in the parent lattice strengthening the alloy. Maximum strengthening occurs when the lattice is subjected to maximum strain, that is, when there are equal numbers of both types of atoms. 50 per cent (a/o) is not necessarily the same as 50 per cent (w/o). Property variations with composition for solid-solution alloys are shown in figure 4.13b.

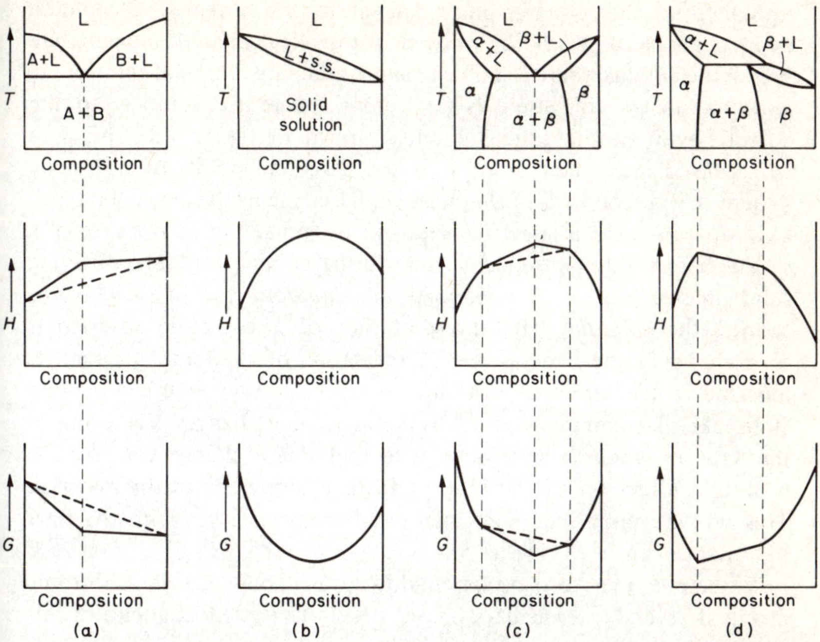

Figure 4.13. *Relationships between alloy composition and hardness (H) and electrical conductivity (G) for: (a) simple eutectic; (b) solid solubility; (c) partial solubility with eutectic; and (d) peritectic*

Figures 4.13c and d show the relationships between properties and composition for the partial solid solubility cases. As the phase diagrams are combination-type diagrams, so the property diagrams are combinations of the former two types.

4.11. Effect of Allotropy on a Phase Diagram

If a substance is allotropic this will have an effect on the shape of phase diagrams for systems involving the substance.

Consider a hypothetical system between two allotropic substances, A and B. Suppose that A is body-centred cubic at low temperatures and face-centred cubic at high temperatures, while B is hexagonal at low temperatures and face-centred cubic at high temperatures. If all factors are favourable it is possible for a complete range of solid solutions to exist between the two at high temperatures. At lower temperatures only partial solid solubility can occur because the substances differ in crystal form. The complete phase diagram for the system A–B could be as in figure 4.14. It will be noticed in this diagram that, immediately below the solidus, there is a phase field containing one solid solution, β, and that below this there is a diagram apparently identical to the eutectic with partial solid solubility shown in figure 4.8. This is a *eutectoid* diagram and point E is the eutectoid point. In a eutectic system it is a case of a single-phase liquid changing during cooling into two separate solid phases. A eutectoid is similar but it is a case of a single-phase solid solution changing during cooling into two differing solid phases. Line CED is termed the *liquidoid* and line CFEGD is termed the *solidoid*. The interpretation of a eutectoid diagram is fundamentally the same as the interpretation of a eutectic diagram. It must be remembered, though, that reactions wholly within the solid state take place more slowly than liquid to solid changes. Variations in the rate of cooling through a eutectoid phase change can exert a profound effect on the structure and final properties of the material. This will be discussed in greater detail in connection with the heat treatment of steels in chapter 5.

Eutectoids may also occur in alloy systems between non-allotropic metals if several intermediate phases occur. There are examples of this in the copper–aluminium and copper–tin systems.

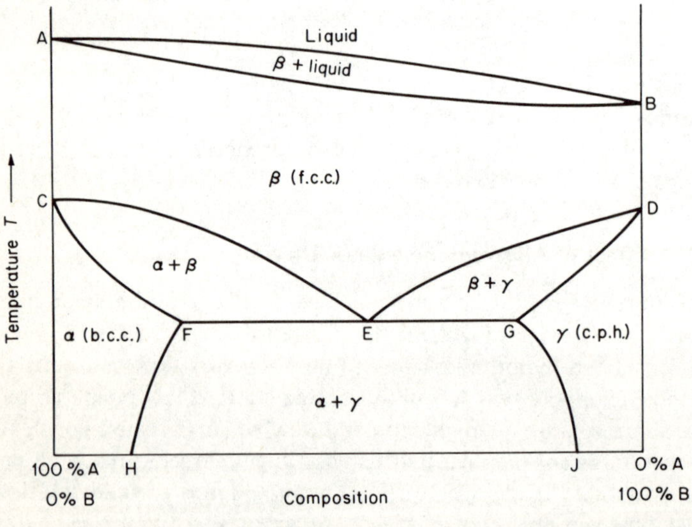

Figure 4.14. *Possible binary phase diagram for two hypothetical allotropic metals*

4.12. Aqueous Solutions

Another type of liquid–solid binary system is that involving the solution of a salt in water. Consider the system water–potassium iodide (KI). The *t–c* diagram for this system, at atmospheric pressure, is shown in figure 4.15. The curve AE represents the freezing-point curve for aqueous solutions of the salt. Similarly to the simple eutectic system discussed in section 4.2, it is crystals of ice which separate out as the liquid solution freezes. The curve EB is a solvus line and indicates the limit of solubility of the salt in water and its variation with temperature. Point E is a eutectic point and is invariant (at constant pressure). This point denotes the lowest temperature at which a solution of the salt in water can exist, is also known as the *cryohydric* point. Freezing of a solution of this composition would take place at constant temperature, resulting in a eutectic mixture of ice and potassium iodide crystals.

A *p–t* diagram for an aqueous-solution system could be drawn as shown in figure 4.16. Curves OA, OB and OC represent the vapour-pressure curves for water and for ice, and the effect of pressure on the freezing point of water, respectively (see section 2.1). The curve EFG represents the vapour-pressure curve of saturated solutions of the salt in water. As the presence of a solute in water lowers the vapour pressure at any temperature, curve EFG will lie below the vapour-pressure curve for water. The vapour pressure of a pure liquid increases with increase in temperature, and, in general, the solubility of a salt in a liquid increases with an increase in temperature. The shape of the vapour-pressure curve for the saturated solution is, therefore, affected by two opposing factors: an increase in temperature tending to increase vapour

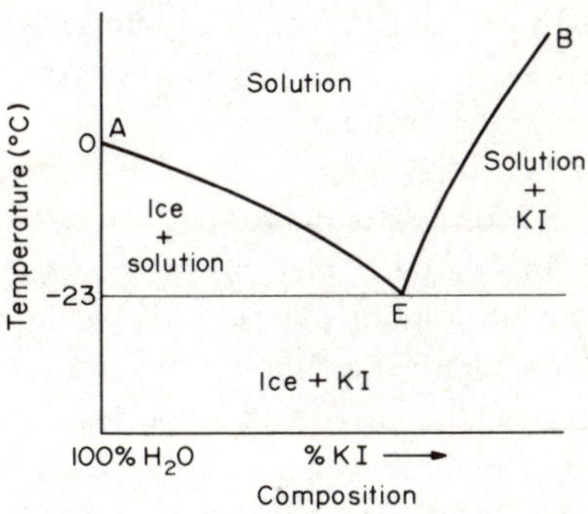

Figure 4.15. *t–c diagram for the water–potassium iodide system*

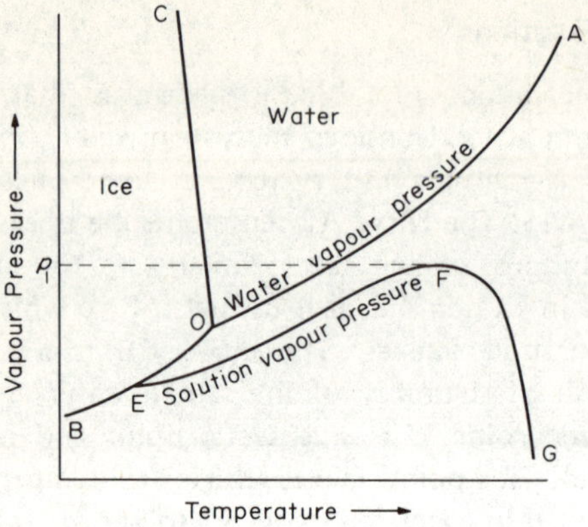

Figure 4.16. *p–t curve for a saturated aqueous solution*

pressure, and a temperature increase tending to raise the solute concentration in a saturated solution, so causing a reduction in vapour pressure. The general form of the solution vapour-pressure curve possesses a maximum as shown by curve EFG in figure 4.16.

Point E in figure 4.16 is the cryohydric point and corresponds to point E in figure 4.15. Point G corresponds to the melting point of the pure salt, generally with a very low vapour pressure. If the pressure p_1 of the maximum point F on the solution vapour-pressure curve is less than atmospheric pressure, it signifies that the saturated solution should never boil at ordinary pressure. This is, in fact, the case with some extremely soluble substances such as sodium hydroxide. If the pressure p_1 is of a higher value than atmospheric pressure then the atmospheric pressure isobar would cut the curve EFG in two places indicating two boiling points. This does happen and there are two boiling points for the saturated solutions of several substances. One substance which shows this behaviour is silver nitrate.

Many salts form hydrates. Sodium chloride may exist in either the anhydrous form, NaCl, or as the dihydrate, $NaCl.2H_2O$. Salt hydrates may possess either incongruent melting points or congruent melting points; the dihydrate of sodium chloride has an incongruent melting point of 0.15°C. The phase diagram for the water–sodium chloride system is, therefore, of the meritectic type (see section 4.8) and is shown in figure 4.17. The dihydrate, $NaCl.2H_2O$, cannot exist above its incongruent melting point of 0.15°C. At this meritectic temperature the reaction which occurs is

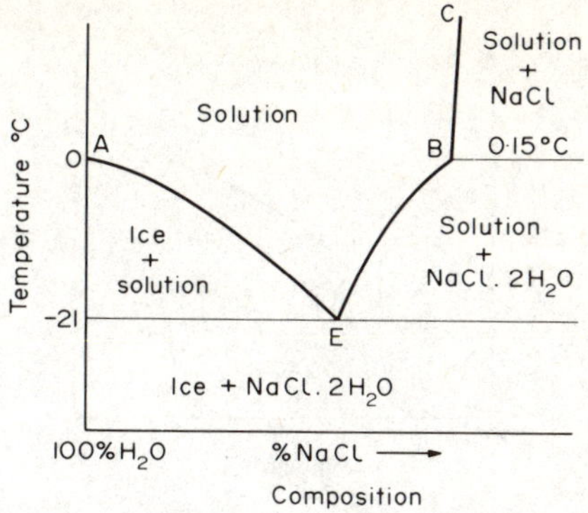

Figure 4.17. *t–c diagram for the water–sodium chloride system*

$$NaCl \,.\, 2H_2O \underset{\text{cooling}}{\overset{\text{heating}}{\rightleftarrows}} NaCl + \text{Solution}$$

In other systems the hydrate or hydrates may possess congruent melting points. In the water–ferric chloride system there are four stable hydrates in addition to the anhydrous salt, Fe_2Cl_6, and each of these possesses a congruent melting point. In the phase diagram (figure 4.18)

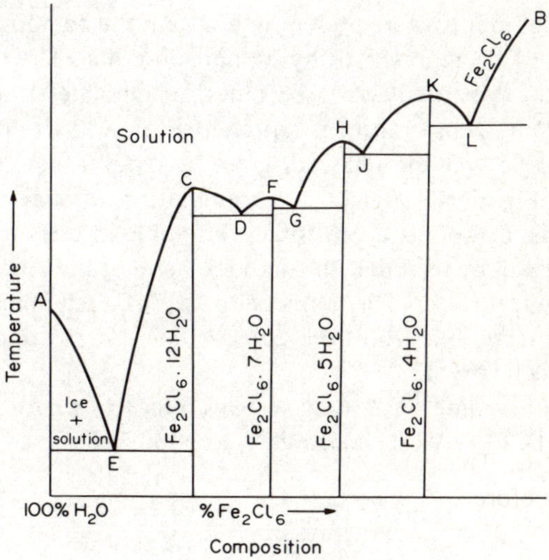

Figure 4.18. *t–c diagram for the water–ferric chloride system*

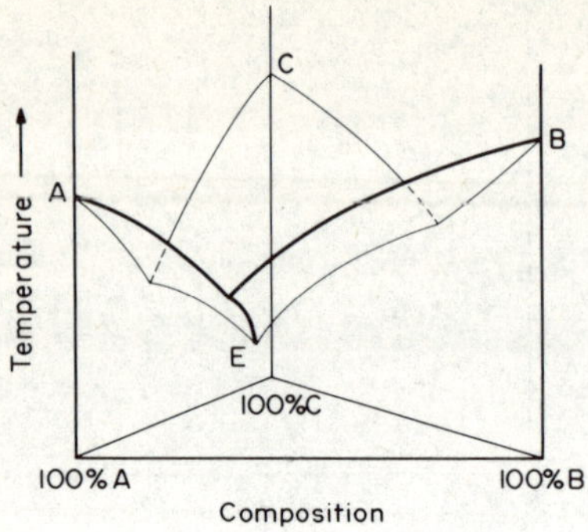

Figure 4.19. *Ternary eutectic diagram*

the points C, F, H, and K represent the melting points of the respective hydrates. Point E is the eutectic or cryohydric point, while points D, G, J, and L are also eutectic points for mixtures of two hydrates. At its congruent melting point a pure hydrate will melt to give a solution in which the salt and water are in the same concentrations as in the solid hydrate.

4.13. Ternary Diagrams

When three components are present in a system the composition of any mixture cannot be represented by a point on a line. The composition abscissa of the binary diagram becomes an equilateral triangle in a ternary diagram. Temperature is represented on an axis orthogonal to the base triangle and the phase diagram becomes a three dimensional solid figure. Figure 4.19 gives a representation of a ternary system where all three components are totally insoluble in one another in the solid state. It will be seen that the liquidus curve of the binary system is now a curved surface. The eutectic point of each binary diagram becomes a eutectic line with three eutectic lines intersecting at point E, a ternary eutectic point.

For further reading on ternary systems refer to *Ternary Equilibrium Diagrams* by D. R. F. West. (Macmillan, London 1965).

5

Real Systems

5.1. Solid Solutions

As mentioned in section 4.10 the crystal lattice of a solid solution will be in a state of strain. Solute atoms will be either larger or smaller than the atoms of solvent, creating either a positive or negative strain in the lattice. Yielding and plastic deformation in metals is due to dislocations being moved through the crystal lattice under the action of an externally applied force. The strained areas of lattice caused by the presence of solute atoms will hinder the movement of dislocations and so the force necessary to move dislocations will be increased. In other words, the presence of solute atoms will cause an increase in the yield strength of the metal. The amount by which the yield stress is raised will depend on the total amount of strain developed in the lattice and this, in turn, is related to the amount of solute present and the magnitude of the difference between the atomic diameters of solvent and solute atoms.

Copper and nickel, both metals crystallising as face-centred cubic, form a complete series of solid solutions. The *t*–*c* phase diagram for the copper–nickel alloy system is of the type shown in figure 4.7a. The atoms of both elements have similar diameters, these differing by only 2.5 per cent. Both the addition of nickel to copper and the addition of copper to nickel result in a raising of the strength and hardness of each pure metal.

While strain in a crystal lattice causes an increase in mechanical strength and hardness, it reduces the electrical and thermal conductivities. Table 5.1 gives strength and electrical resistivity data for alloys of copper and nickel.

Table 5.1. Solution strengthening: properties of annealed copper–nickel solid solutions

Composition		Hardness	Tensile Strength	Elongation	Resistivity
Cu%	Ni%	V.P.N.	MN/m^2	%	Ω m
100	–	60	210	65	1.67×10^{-8}
95	5	65	230	50	2.8×10^{-8}
90	10	70	250	45	5.1×10^{-8}
80	20	75	315	45	1.1×10^{-7}
70	30	80	370	45	2.8×10^{-7}
60	40	90	430	45	5.5×10^{-7}
30	70	120	520	45	3.8×10^{-7}
–	100	95	450	50	6.84×10^{-8}

An interesting comparison is obtained by examining the effect of small additions of solute elements with a large difference between the effective diameters of solvent and solute atoms. The strength and hardness of high-purity aluminium, and some other aluminium base materials is given in table 5.2. Commercial-purity aluminium contains about 0.5 per cent of iron and about 0.3 per cent silicon in solution. The aluminium–magnesium alloys quoted are made from commercial purity aluminium with additions of high-purity magnesium. The diameter of an iron atom is about 13 per cent smaller than that of aluminium. Silicon atoms are about 11 per cent smaller than aluminium, while magnesium atoms are about 11.5 per cent larger in size than those of aluminium.

It will be seen from tables 5.1 and 5.2 that a small addition of an element with a large size difference will have a far greater effect on the

Table 5.2. Solution strengthening: properties of annealed aluminium materials

Composition	Tensile Strength MN/m^2
High-purity aluminium	
99.99% Al	45
Commercial purity	
Al + 0.5% Fe + 0.3% Si	90
Al + 2% Mg	210
Al + 3½% Mg	230
Al + 5% Mg	280

properties of solid solutions than a similar amount of an element with a small size difference.

5.2. Eutectics

There are numerous real systems containing eutectics. One of the features of a eutectic system is that the eutectic mixture melts at a temperature below the melting points of the constituents.

The phase diagram for aluminium, melting point 659°C, and silicon, melting point 1430°C, shows a eutectic with limited solid solubility (similar to figure 4.8). The eutectic occurs at 11.7 per cent of silicon and the eutectic temperature is 577°C. Alloys of aluminium and silicon with compositions at or near the eutectic composition are eminently suitable as casting alloys for several reasons. A eutectic alloy not only has a comparatively low melting point but it also freezes at a constant temperature (or if the alloy composition is close to that of the eutectic the alloy will freeze over a narrow range of temperature). This is a useful property for casting alloys as it reduces segregation effects to a minimum. The aluminium–silicon alloys are extremely good alloys for both sand and die casting.†

The phase diagram for lead and tin is also a eutectic type with partial solid solubility. The eutectic alloy, which contains 38 per cent of lead, melts at 183°C. Because of their low melting temperatures and good 'wetting' characteristics lead–tin alloys are used as solders. For jointing tin plate and for joints in electrical wiring the eutectic alloy is used, since it freezes at one temperature. This composition is known as tinman's solder. Plumber's solder, on the other hand, contains 66 per cent of lead and solidifies over a wide freezing range. For this composition alloy the liquidus temperature is 270°C and the solidus temperature is 183°C. The wide freezing range allows plumbers to make wiped joints when joining lead pipes.

When antimony is added to lead–tin alloys there is a ternary eutectic formed. Ternary lead–tin–antimony alloys of various compositions are used for the casting of type and white metal bearings, both applications where an alloy of low melting point is desirable.

†Aluminium–silicon alloys are 'modified' with a small addition of sodium. This refines the grain structure of the eutectic mixture and also affects the phase diagram slightly moving the eutectic point to 14 per cent silicon at 564°C. The silicon content of an Al–Si casting alloy should be less than the eutectic value as the presence of primary silicon crystals would render the alloy brittle.

The melting point of a substance is often lowered by alloying even in those systems with no eutectic. The addition of zinc to copper, to make brasses, causes a lowering of melting point (section 5.5). A brass containing 50 per cent of zinc melts at about 860°C, compared with 1083°C for the melting point of copper. The alloy of this composition is used for making brazed joints. Addition of silver to brass reduces the melting point of the alloy still further and gives an increase in strength. These ternary alloys form the basis of the silver solders (hard solders) used extensively for metal joining.

5.3. Precipitation Hardening

Precipitation hardening is a useful process for strengthening certain alloy compositions. The process is only possible because of the creation, by suitable heat treatment, of a metastable condition. The alloy systems in which this process may be possible are those in which there is partial solid solubility. A good example is the precipitation hardening of aluminium–copper alloys. The aluminium-rich end of the aluminium–copper phase diagram is shown in figure 5.1a. Within this system the alloy compositions that respond best to precipitation hardening are those containing between 4 per cent and 5.5 per cent of copper. It will be noticed in figure 5.1a that the maximum solubility of copper in aluminium is 5.7 per cent (by weight) at the eutectic temperature 548°C but is only 0.2 per cent (by weight) at low temperatures. When an alloy containing, say, 5 per cent of copper is heated to 550°C all the copper present will be held in solid solution in the aluminium lattice. If the alloy is allowed to cool slowly from this temperature, equilibrium conditions will be established as the solvus line is crossed and the second phase will be precipitated from saturated solid solution. The second phase in this case is a compound, $CuAl_2$, known as θ phase. After slow cooling to room temperature the equilibrium structure will consist of a coarse precipitate of θ phase in a dilute solid solution of copper in aluminium. If, however, the alloy were to be rapidly cooled from 550°C by quenching in water, the whole of the copper content would be retained in solid solution within the aluminium. Such a solution, being highly supersaturated with dissolved copper, is a non-equilibrium phase and hence is metastable. It will possess a tendency to change into the stable structure of dilute solution plus θ precipitate. The concentration of copper in θ phase $CuAl_2$, is 53.5 per cent by weight or 33.3 atomic per cent. The

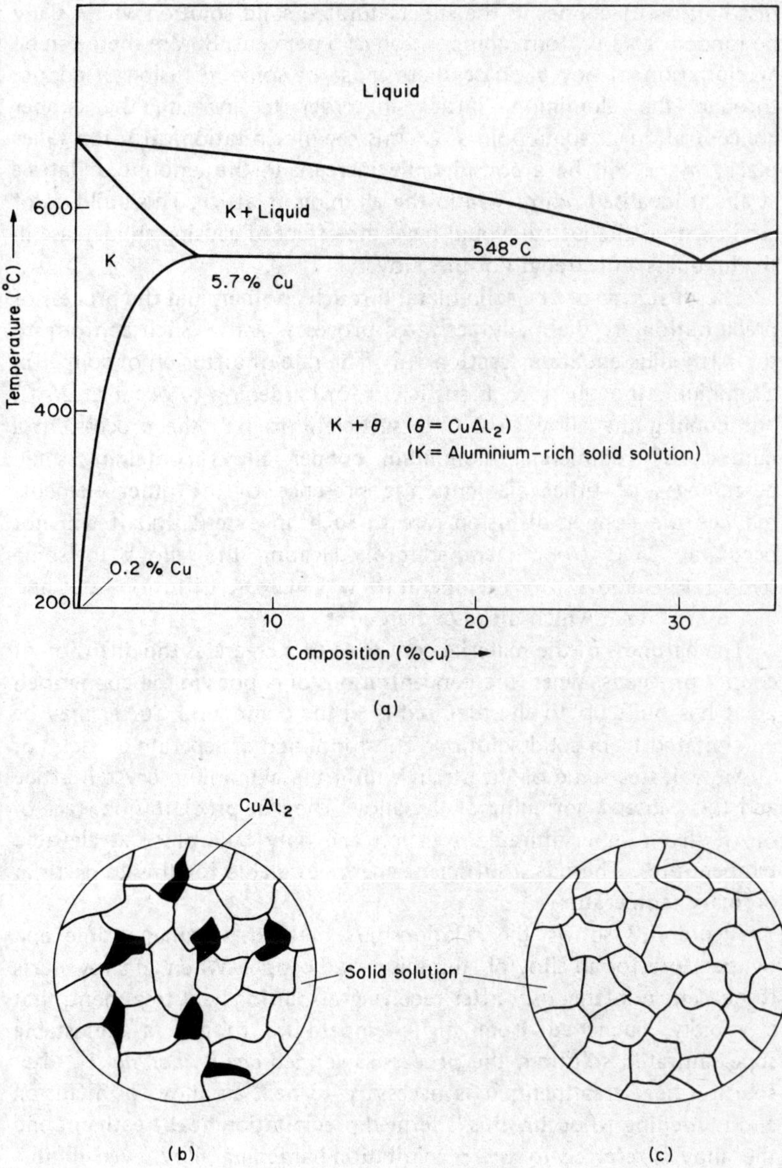

Figure 5.1. *(a) Aluminium-rich end of aluminium–copper phase diagram. (b) 5% Cu alloy slow-cooled from 550°C showing $CuAl_2$ precipitate. (c) 5% Cu alloy rapidly cooled from 550°C - supersaturated solid solution*

distribution of copper in the supersaturated solid solution will initially be random at a uniform composition of 5 per cent. Before there can be precipitation of any θ phase there must be some diffusion of copper through the aluminium lattice in order to increase the copper concentration at some points. As this pre-precipitation diffusion takes place, there will be a considerable increase in the amount of lattice strain at localised points within the aluminium lattice. This build-up of strain within the lattice would have the effect of causing an increase in the hardness and strength of the alloy.

The diffusion of one solid metal through another, and the process of precipitation are thermally activated processes and as such conform to the Arrhenius equation (section 1.6). The rate of diffusion of copper in aluminium, though slow, is sufficient for hardening to occur at 25°C, but cooling the alloy to 0°C is sufficient to halt the process. In a number of commercial aluminium–copper alloys containing small percentages of other elements the presence of the other elements reduces the copper diffusion rate to such an extent that it will not occur at 25°C (room temperature). Heating the alloys to some temperature above room temperature will increase diffusion rates and hence the rate at which the alloy hardens.

The hardness of the material continues to increase as the diffusion of copper proceeds. When the concentrations of copper in the copper-rich areas has built up to the required level the compound $CuAl_2$ may be precipitated from solid solution. The formation of separate particles of $CuAl_2$ releases some of the strain within the aluminium crystal lattice and this causes a softening of the alloy. The true precipitation stage of $CuAl_2$ from supersaturated solution can only take place at elevated temperatures. There is insufficient energy available for this to occur at ordinary temperatures.

Figure 5.2 shows the relationships between hardness, time and temperature for an alloy of aluminium and copper. When an alloy starts to harden spontaneously after receiving a solution heat treatment, that is rapidly quenched from high temperature to give a metastable supersaturated solution, the process is termed *age hardening*. If, after solution heat treatment, it is necessary to heat an alloy for diffusion and hardening to occur, this is termed precipitation heat treatment and the alloy is referred to as a precipitation-hardening alloy, even though the treatment is ceased before true precipitation and softening can occur. The softening at high temperature is termed *over-ageing*.

Any binary system for which the phase diagram shows partial solid

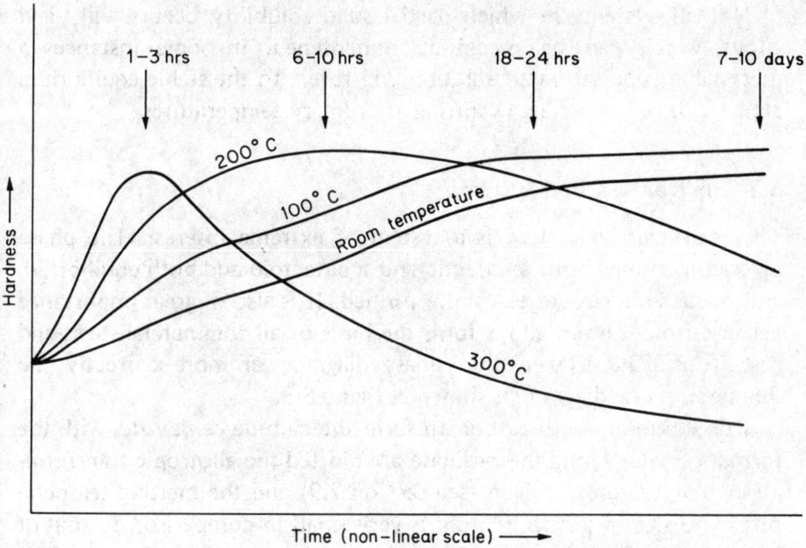

Figure 5.2. *Hardness–time relationships for aluminium–copper age hardening alloy*

solubility may possess alloy compositions which will respond to a precipitation-hardening process. There are other alloys based on aluminium which may be strengthened in this manner. Nickel is partially soluble in aluminium and alloys containing both nickel and copper are used extensively. The presence of nickel reduces diffusion rates within aluminium and in consequence Al–Cu–Ni alloys can be heated to higher temperatures than straight Al–Cu alloys before over-ageing occurs. These alloys are used for applications in which retention of strength at elevated temperatures is a requirement. They are used in the construction of the Concorde supersonic air transport, where frictional heating occurs at high airspeeds, and for the cylinder blocks and heads of internal-combustion engines. Magnesium and silicon form a compound, Mg_2Si, when added in the correct proportions and the phase diagram between this compound and aluminium shows partial solid solubility. Al–Mg–Si alloys will respond to precipitation hardening.

Other systems in which this process is used to strengthen commercial alloys include Cu–Be, Cu–Cr, and certain alloy steels rich in nickel – maraging steels.

Not all systems in which partial solid solubility occurs will yield alloys which can be precipitation hardened. In some instances a metastable supersaturated solution will revert to the stable equilibrium state, with full softening occurring at ordinary temperatures.

5.4 The Iron–Carbon System

The iron–carbon system is a system of extreme interest. The phase diagram contains both a eutectic and a eutectoid and both equilibrium and metastable structures can be formed. It is also of great importance because iron–carbon alloys form the basis of all commercial steels and cast irons. The iron–carbon phase diagram, or more correctly the Fe–Fe_3C phase diagram, is shown in figure 5.3.

Iron combines with carbon to form the carbide *cementite*, with the formula Fe_3C. Along the ordinate are plotted the allotropic transformation temperatures of iron (see section 2.9) and the melting temperature. The size of a carbon atom is very small in comparison to that of iron and solid solutions of the interstitial type are formed. The solubility of carbon in α iron (body-centred cubic) is extremely limited but the solubility of carbon in γ iron (face-centred cubic) is consider-

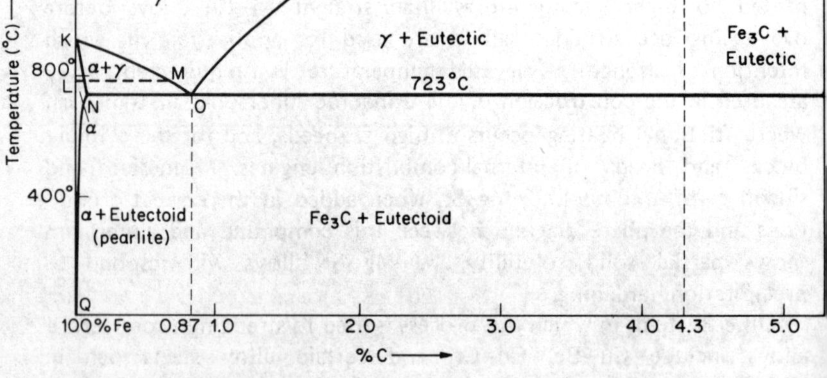

Figure 5.3. *Fe–Fe₃C phase diagram*

ably greater reaching a maximum limit of 1.7 per cent carbon at 1130°C. Names have been assigned to the various phases within the $Fe-Fe_3C$ system, as follows: the body-centred cubic phases, α and δ, are termed *ferrite*; the face-centred cubic phase, γ, is termed *austenite*; and the eutectoid mixture of α and cementite is known as *pearlite*.

The full $Fe-Fe_3C$ phase diagram may appear at first sight to be extremely complex but it can be divided into sections which, in themselves, are straightforward. Steels are basically alloys of iron and carbon containing up to 1.5 per cent of carbon. Therefore for the consideration of steels, and in particular their heat treatment, it is convenient to consider only that portion of the diagram up to a carbon content of 1.5 per cent and up to a temperature of 1000°C (see figure 5.4).

It will be seen that ferrite cannot hold carbon in solid solution to any great extent, the limits being 0.04 per cent of carbon at 723°C and 0.006 per cent of carbon at 200°C. Austenite, however, can hold a considerable amount of carbon in solid solution, ranging from 0.87 per cent at 723°C to 1.7 per cent at 1130°C. The eutectoid point occurs at a temperature of 723°C and at a carbon content of 0.87 per cent. The terms hypoeutectoid and hypereutectoid are used to denote steels that contain less carbon than, and more carbon than, the eutectoid composition, respectively.

The presence of carbon depresses the $\alpha-\gamma$ transformation temperature of iron. Line KMO in the figures denotes this transformation temperature, and its dependence on composition. Lines OF and QN are solvus lines and denote the maximum solubility limits of carbon in γ and α iron respectively. Point O is the eutectoid, or pearlite, point. The line LMOP indicates the Curie temperature, at which loss of magnetism occurs.

If a sample of a steel is heated or cooled, and accurate measurements are taken, thermal arrest points will be noted, corresponding to the phase-transformation lines (and Curie temperature) on the phase diagram. The phase line NOP is known as the A_1 transformation, the Curie temperature, LM as the A_2 transformation, line KMO as the A_3 transformation, and the line OF as the A_{cm} transformation. These arrest points, or transformation temperatures, are also known as the *critical points*, or *critical temperatures*, for the steel. The eutectoid temperature, A_1, is known as the lower critical temperature, and the α to γ transformation, line KMO, is known as the upper critical temperature. If a steel is heated or cooled very slowly, so that

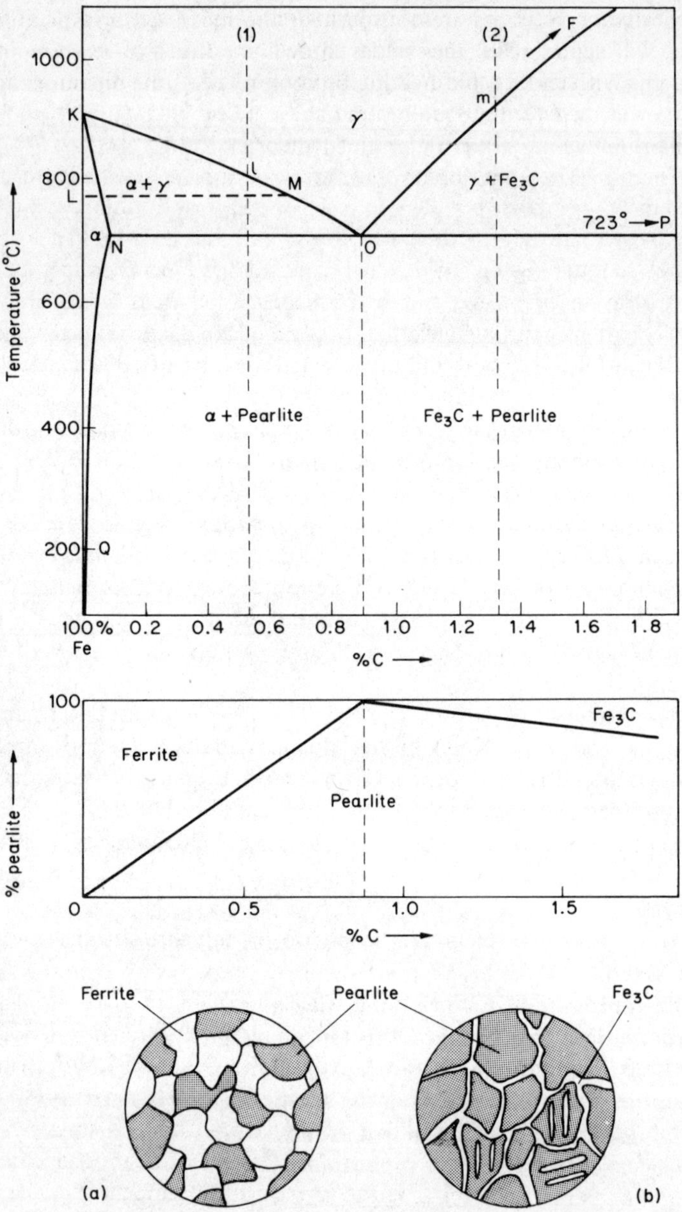

Figure 5.4. *Steel portion of the Fe–Fe_3C diagram: (a) microstructure of hypoeutectoid steel (1); (b) microstructure of hypereutectoid steel (2)*

equilibrium conditions are approached, the measured arrest temperatures will agree with the values shown on the iron–carbon phase diagram. With more rapid heating or cooling rates, the measured arrest points will differ from equilibrium values. They will be higher than the equilibrium values when determined during heating, and lower when determined during cooling. Values measured during heating are written as Ac_1, Ac_2, and Ac_3, while values determined during cooling are written as Ar_1, Ar_2, and Ar_3.†

Let us now consider the changes that occur during the cooling of steels of various compositions. Refer to figure 5.4 and consider first the cooling of a hypoeutectoid steel of composition (1) At a high temperature the steel structure will be composed of homogeneous crystals of austenite solid solution. On cooling to the upper critical temperature, point l on the diagram, austenite will begin to transform into ferrite. The ferrite can hold very little carbon in solid solution and so the remaining austenite becomes enriched in dissolved carbon. As the temperature falls, more ferrite is formed, and the composition of the remaining austenite increases in carbon content, following the line KMO. When the lower critical temperature is reached, the austenite, which is now of eutectoid composition, transforms into the eutectoid mixture pearlite, a mixture composed of alternate layers of ferrite and cementite.

For a hypereutectoid steel of composition (2) the homogeneous austenite structure which exists at high temperatures will begin to change when the temperature has fallen to a point, m, on the solvus line OF. This is the saturation limit for dissolved carbon in austenite and on cooling below the temperature of point m, excess carbon precipitates from solid solution in the form of cementite. The cementite appears in the microstructure as a network around the austenite crystals, and also in the form of needles within the austenite crystal grains. This latter type is termed Widman–Statten type precipitation. The carbon content of the austenite reduces with further cooling and when the lower critical temperature is reached all remaining austenite, which is now of eutectoid composition, transforms into pearlite. The presence of cementite in the form of needles, or as a boundary network, renders the steel brittle, and heat treatment is necessary to put the steel into a suitable condition for many applications.

†The symbol *A* signifies arrest (French *arrière*). The suffixes *c* and *r* derive from the French *chauffage*: heating; *refroidissement*: cooling.

Ferrite is a comparatively soft and ductile constituent possessing a tensile strength of about 280 MN/m^2. The tensile strength of pearlite formed by slow cooling from the austenitic range is about 700 MN/m^2, but its ductility is very much less than that of ferrite. It must be emphasised that the properties quoted only apply to slowly cooled steels. An increase in the rate of cooling through the critical temperature range will alter the structure, and hence the properties, of any steel. When the cooling rate is increased, there is some undercooling of austenite to below the equilibrium transformation temperatures. Once the phase change from undercooled austenite to pearlite commences it takes place very rapidly, resulting in very fine lamellae of ferrite and cementite. The hardness and strength of pearlite is dependent on the interlamellar spacing, and very fine pearlite formed by rapid cooling may have tensile strengths of the order of 1300 MN/m^2.

If a steel is cooled extremely rapidly there will be insufficient time allowed for austenite to decompose into pearlite and, instead, the austenite changes into a metastable body-centred lattice with all the carbon trapped in interstitial solid solution. Ferrite can theoretically hold virtually no carbon in solution. The rapidly cooled structure that is formed is highly strained and distorted by the large amount of dissolved carbon into a body-centred tetragonal lattice. This constituent is termed *martensite*, which is extremely hard and brittle. The hardness of martensite depends on the carbon content, and is greatest (that is the greatest degree of lattice strain) in high carbon steels. Under the microscope, martensite appears as a series of fine needle-like (acicular) crystals. A martensitic structure can be formed by rapidly quenching a heated steel, from the austenitic state, into water or oil. This is the treatment termed hardening.

Martensite is a non-equilibrium phase which does not appear in the iron–carbon phase diagram. In order to show the influence of cooling rates, that is time, on the transformation of austenite, another type of diagram is necessary. This is the time–temperature–transformation, or T–T–T diagram. T–T–T diagrams are sometimes known as 'S curves' because of their general shape.

A typical T–T–T diagram for plain carbon steel is shown in figure 5.5. It will be seen that a slow cooling rate will lead to the formation of coarse pearlite, with little undercooling of austenite, while a faster cooling rate will give a greater amount of undercooling and the formation of fine pearlite. If the critical cooling velocity is exceeded,

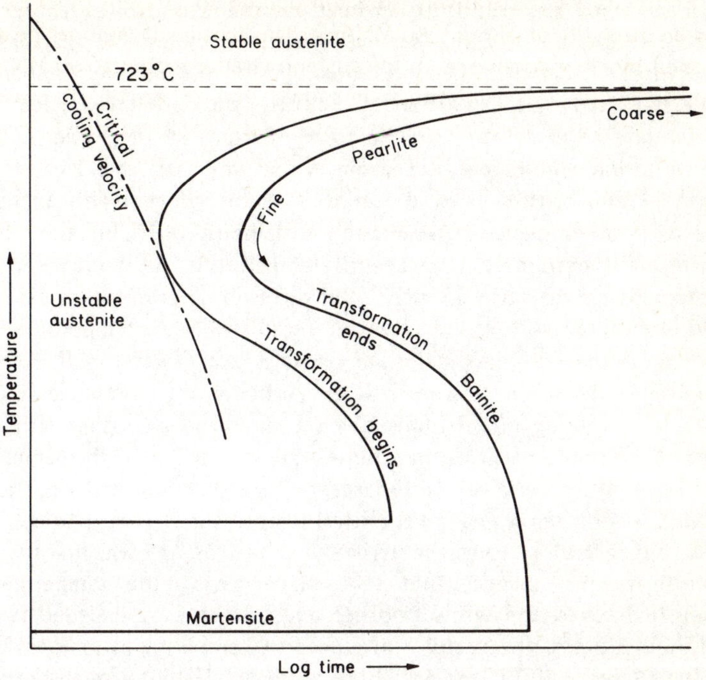

Figure 5.5. *T–T–T diagram (S-curve) for a plain carbon steel*

the non-equilibrium phase, martensite, will be formed. *Bainite*, another non-equilibrium phase, is a finely divided dispersion of carbide particles in ferrite and is formed by the isothermal transformation of under-cooled austenite, namely, by the rapid quenching of the steel to a temperature below the nose of the T–T–T curve and then maintaining constant temperature until transformation is complete.

The position of the nose of the curve, and hence the value of the critical cooling velocity, is not constant for all steels. An increase in the carbon content of the steel, or an increase in the content of other alloying elements, will reduce the value of the critical cooling velocity, that is move the T–T–T curve toward the right. The rate of cooling possible by quenching a steel in water is about equal to the critical cooling velocity of a plain carbon steel containing 0.3 per cent of carbon. Consequently, it is impracticable to harden, by quenching, plain carbon steels with a lesser carbon content than this.

Martensite, although very hard, is also extremely brittle and a hardened steel requires a further heat treatment, known as *tempering*, before it can be put into service. When the metastable martensitic structure is heated it becomes possible for the carbon trapped in supersaturated solid solution to diffuse through the lattice and precipitate from solution in the form of iron carbide particles. This precipitation will relieve the strain within the lattice and cause the hardness and brittleness of the material to be reduced. This diffusion process can commence at temperatures of about 200°C, but the rate of diffusion is extremely slow at this temperature. An increase in the temperature will cause an increase in diffusion and precipitation rates and, therefore, increase the extent of the softening. At temperatures up to 450°C the carbide precipitate particles are much too fine to be resolved under the optical microscope, although their presence may be detected by using more sophisticated techniques. At higher temperatures the carbide particles increase in size, and at 700°C the cementite coalesces into a series of fairly large, and roughly spheroidal particles. (700°C is just below the lower critical temperature.) This gives rise to a soft, but incredibly tough, material. When microscopy was first used to investigate the changes that take place during the tempering of quenched steels, the terms *troostite* and *sorbite* were assigned to the distinctive types of structures produced by tempering at temperatures in the region of 400°C and 500°C, respectively. These terms still remain in use although the structures could be more properly described as *tempered martensite*. Tempering temperatures and times have to be fairly accurately controlled in order to produce the desired properties in the material.

Alloying elements are frequently added to steels to bring about improvements in properties and the added elements will have an effect on the phase relationships. Any alloying element will influence the α–γ transformation temperature and may either increase or decrease this (figure 5.6). Chromium, tungsten and silicon are some of the elements that raise the transformation temperature, while nickel and manganese are elements that lower it. There will be similar alterations in the critical temperatures of steels that contain alloying elements in addition to carbon.

When a large amount of alloying element is present a phase transformation may be eliminated and the steel becomes either wholly ferritic or wholly austenitic. It should be noted that if the α to γ phase change is absent then it is not possible to create the metastable martensite phase by heat treatment.

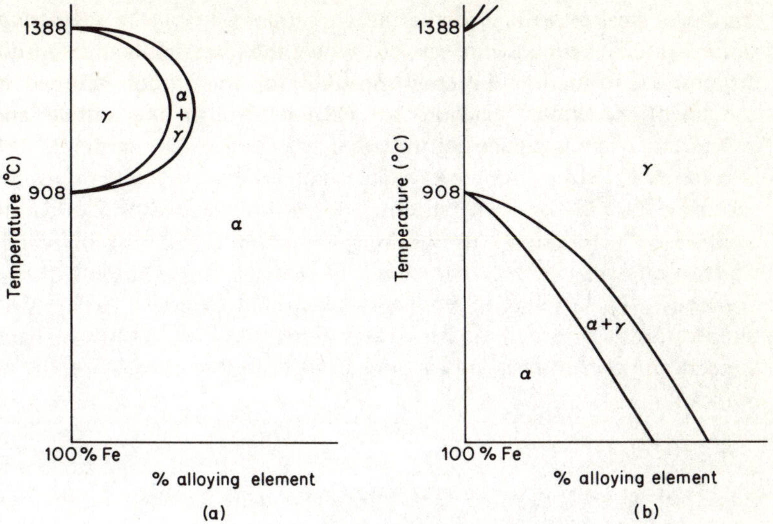

Figure 5.6. *Effect of alloying addition on allotropic transformation temperatures of iron: (a) ferrite stabilising (γ loop) element – (Cr, W, Si); (b) austenite stabilising element – (Ni, Mn)*

Many alloying elements form stable carbides. Some, such as manganese carbide, Mn_3C, are associated with cementite, Fe_3C, as a single phase, while others, notably carbides of chromium and tungsten form as separate phases in the structure.

Cementite is itself only metastable but in iron–carbon materials containing less than 1.5 per cent of carbon, that is steels, the stable phase, graphite does not occur.† In cast irons, however, which possess carbon contents within the range 1.5 to 5 per cent, graphite frequently appears in the structure. There are several factors that can affect the type of structure found in a cast iron. Fairly slow rates of solidification, such as are experienced in sand casting, will allow an approach to equilibrium conditions and there will be a tendency for the casting to be a grey iron; that is, stable graphite present in the structure causes a fracture surface to have a grey appearance. Chill casting, which gives a faster rate of solidification, will tend to give cementite in the iron

†The transformation from metastable cementite to stable graphite is catalysed by nickel and silicon, consequently the carbon content is restricted to less than 0.5 per cent in silicon or nickel alloy steels.

structure. An iron with all the carbon in the combined form is termed a white iron. The presence of the elements silicon or nickel in iron will promote the formation of graphite in the structure.

Cementite, although technically metastable, will remain indefinitely in an iron structure under normal conditions, but will slowly transform into the stable states iron and graphite if maintained at temperatures in excess of 600°C for long periods. As graphite is of low density in comparison with cementite, the high temperature graphitisation of a white iron will be accompanied by an increase in dimensions. Consequently, cast-iron components required for service at elevated temperatures must possess wholly graphitic structures. High-temperature growth-resistant cast irons are alloy irons containing silicon or nickel.

5.5. Systems with Intermediate Phases

Many well known alloy systems contain numerous intermediate phases. As examples of this, portions of three-phase diagrams, Cu–Zn, Cu–Sn and Cu–Al are shown in figure 5.7.

The three examples shown appear very complex but close examination will reveal that each of the diagrams is composed of the comparatively simple sections discussed in chapter 4, namely areas showing partial solid solubility, peritectic and meritectic changes etc., and interpretation of the diagrams follows the rules quoted in sections 3.4 and 4.3.

There are a few specific points relating to these diagrams which are worth mentioning. In the copper–zinc system the phases β and β' appear. Both β and β' possess body-centred cubic crystal structures. At a high temperature the β phase is a random substitutional solid solution. On cooling through 450°C β transforms into β', which is an ordered solid solution. This latter phase could be more properly considered as an intermetallic compound CuZn.

In the portion of the copper–tin diagram shown there are several eutectoid transformations. The room-temperature structure of a bronze contains the $\alpha + \delta$ eutectoid mixture but δ phase is technically metastable. The true equilibrium structure should contain $\alpha + \epsilon$ eutectoid mixture formed by the transformation of δ phase at 350°C. The rate of this transformation is very slow at 350°C and ϵ is not formed in normal circumstances.

Another feature of the Cu–Sn diagram is the wide separation

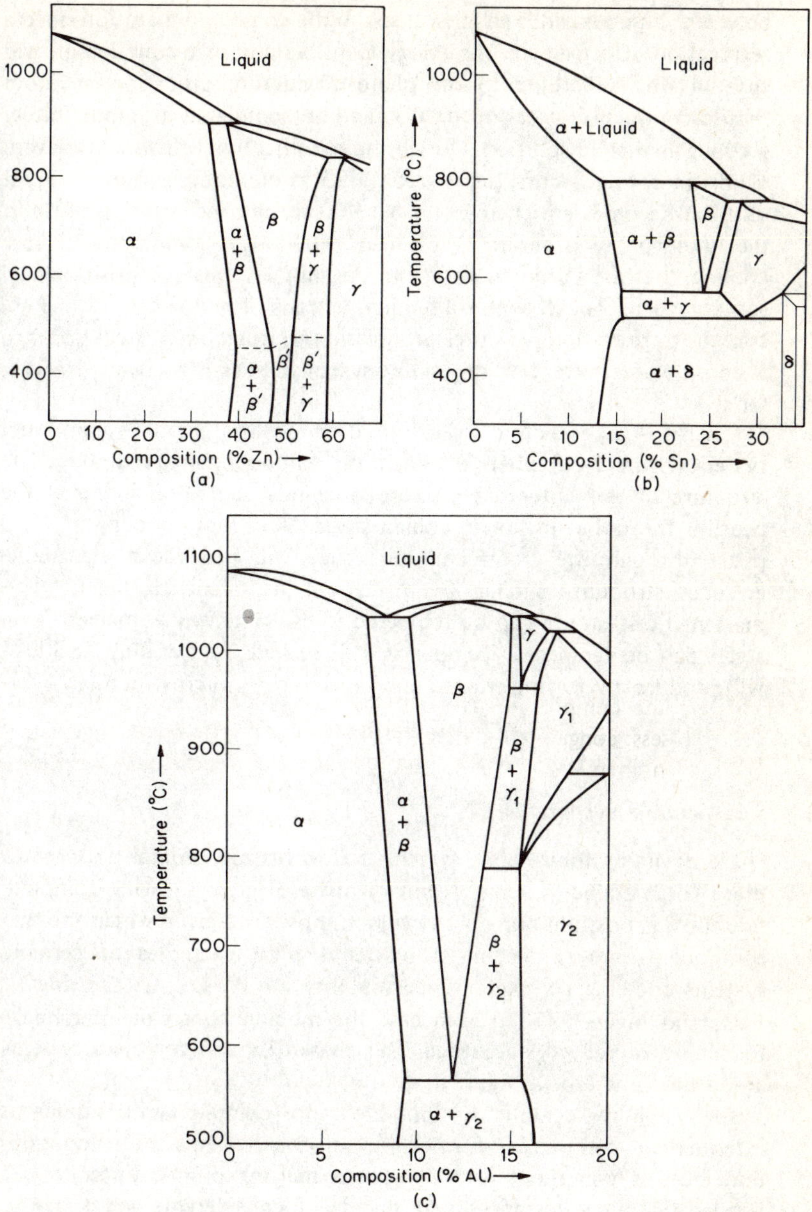

Figure 5.7. *Binary phase diagrams for copper alloys: (a) copper–zinc; (b) copper–tin; (c) copper–aluminium*

between liquidus and solidus lines. With a wide separation severe segregation effects occur in castings. Consider a bronze containing 7 per cent of tin. According to the phase diagram solidification of liquid should commence at about 1030°C and be completed at about 900°C forming an α solid solution. During the solidification of this alloy severe dendritic coring occurs (see section 4.5.) as the composition of liquid solution becomes enriched in tin. At 900°C the concentration of tin in the liquid phase is about 21 per cent. Even with comparatively slow cooling the last liquid to freeze in an alloy of this composition will solidify as β. The β will subsequently transform through γ to δ on cooling to room temperature. Segregation of constituents will occur to a considerable extent in any alloy system possessing a wide freezing range.

There is a eutectoid transformation in the copper–aluminium system (figure 5.7c) where β transforms on cooling into α and γ_2. The structure of this eutectoid mixture is lamellar and very similar to the pearlite formation in slowly cooled steels. Very rapid cooling of the β phase through the transformation range will produce a metastable acicular structure similar to martensite in quenched steels. This martensitic structure can be tempered in the same way as martensite in steels can be tempered. Tempering a quenched Cu–Al alloy at 500°C will produce a very fine grained and tough $\alpha + \gamma_2$ microstructure.

5.6. Ceramic Systems

Phase diagrams for ceramic systems follow broadly similar patterns to those for metallic systems. A binary phase diagram showing complete solubility is possible but only occurs in those systems in which the two components are similar in structure. Typical examples of ceramic systems showing complete solid solubility are Al_2O_3–Cr_2O_3, MgO–FeO, and MgO–NiO. In each case the metallic ions can interchange forming a mixed crystal lattice. The phase diagrams for these systems are of the same type as figure 4.7a.

Many binary ceramic systems are more complex with numerous intermediate compounds formed. As in other systems the compounds may possess congruent or incongruent melting points. Eutectics are formed in many instances and this is of considerable advantage in formulating slag compositions so that a furnace slag will be molten at a relatively low temperature. One major difference between metallic systems and ceramic systems is in the rate at which equilibrium is achieved. Liquid metals at temperatures just above the liquidus are of low viscosity in comparison with most ceramics under similar conditions. Even with extremely rapid solidification metal atoms can move into their equilibrium positions and form crystal lattices. In a liquid ceramic the molecules and complex ions of the substances, by virtue of their bulk, are sluggish in movement. Under these conditions it is often difficult for equilibrium to be established. During the cooling of liquid silica the slow moving basic groupings of $SiO_4{}^{4-}$ ions cannot readily diffuse into the equilibrium positions to create the stable crystal lattice structure of β quartz. With very slow cooling the liquid silica crystallises into the metastable cristobalite form, but with more rapid cooling liquid silica forms into a glass. A glass is essentially a supercooled liquid of extremely high viscosity. It does not possess the regular crystalline structure characteristic of the true solid. The glass state can be formed in many ceramic materials, particularly in some of the complex silicate materials. Some of these complex silicates have optical transparency and form the range of materials known commercially as glass.

The glass state is a metastable state and there will be a tendency for the material to crystallise or vitrify. As the crystalline state is more dense than glass the process of slow devitrification will lead to the development of tensile stresses within the material and this may cause cracks to appear.

Examples of phase diagrams involving ceramics are shown in figure 5.8.

5.7. Polymer Systems

Polymer phases cannot be treated in the same way as other materials since individual polymer molecules are extremely large and may contain between 10^4 and 10^5 carbon atoms. Also the individual molecules that make up the polymer are not identical but are of variable molecular weight. The melting point of a polymer is not a constant of the material in the same way that the melting point of a pure metal or a ceramic is constant but is a function of the average molecular weight. When in the molten state many polymers begin to decompose or degrade. Also, if two polymers are mixed together in the molten state there is a strong possibility of chemical interaction taking place between them. Many linear polymers are completely immiscible in one another in the solid state and while they can be blended together when cold by a

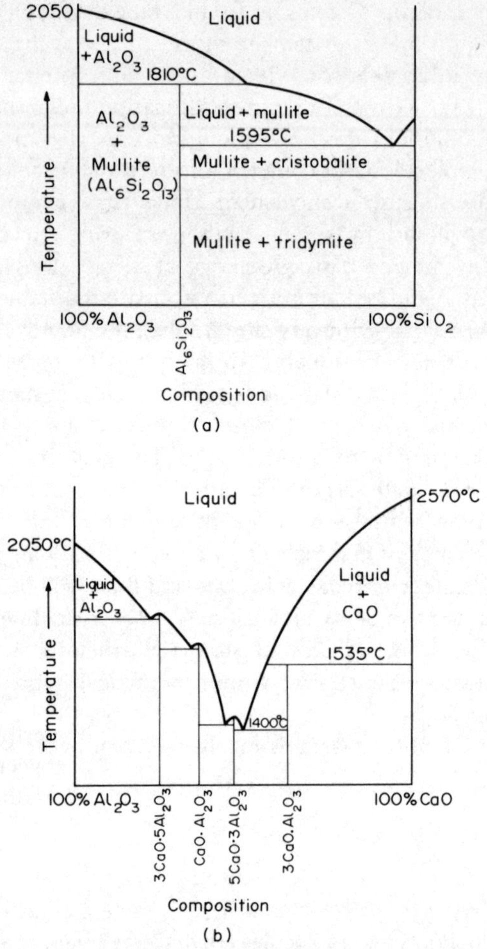

Figure 5.8. *(a) Al_2O_3,–SiO_2 phase diagram. (b) Al_2O_3–CaO phase diagram*

mechanical mixing process, subsequent separation into the respective components may take place at moulding temperatures.

If two compatible monomers are copolymerised in varying proportions it is found that the melting temperatures of the various copolymer compositions follow the pattern in figure 5.9. There is a minimum melting point value for one particular copolymer composition, but figure 5.9. is not a phase diagram in the true sense. Each point on the

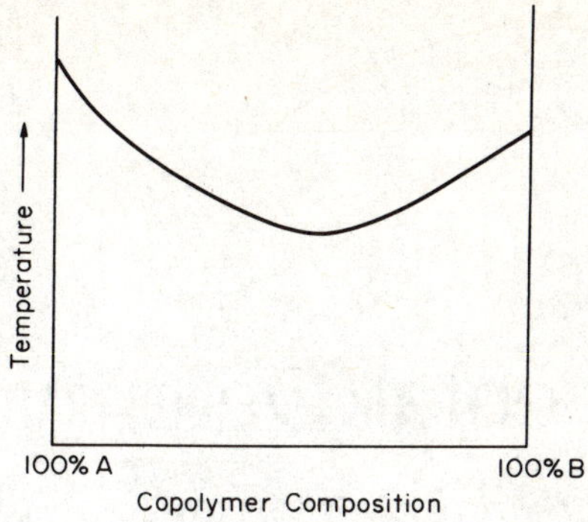

Figure 5.9. *Melting point curve for random copolymers of two components A and B*

melting point curve indicates the melting point of a specific copolymer, and each copolymer is really a new component. In the true binary phase diagram any intermediate composition is a mixture of the two components and not a component in its own right.

6

Experimental Determination of Phase Diagrams

6.1. Introduction

Numerous experimental techniques may be used to establish the boundaries within a phase diagram. In the following paragraphs a number of simple but effective methods, which may be used as the basis of a student laboratory programme are discussed. Other techniques involve the use of sophisticated equipment and are used mainly by research workers. An alternative name for phase diagrams is equilibrium diagrams and the attainment of equilibrium is an important part of many of the practical methods. When a dynamic method is used, as for example in the thermal analysis of a steel to determine thermal arrest points, the results obtained will not be true equilibrium transformation temperatures. Arrest points noted during heating will lie above the equilibrium values and those noted during cooling will be lower than equilibrium values. If the rates of heating and cooling in the test were both similar and fairly slow the mean of the heating and cooling values will approximate to the true equilibrium values.

6.2. Freezing Point Determination

A pure liquid freezes at a constant temperature and its freezing point may be determined either by direct observation of the temperature at which freezing takes place, or by plotting a time–temperature curve during the cooling of the liquid. It is important that the rate of cooling in each case be slow, otherwise undercooling of the liquid may occur. The first method suggested is particularly suitable for those liquids that freeze at temperatures fairly close to ordinary temperatures. The liquid can then be contained in a glass tube into which is inserted a bung holding a thermometer and glass stirrer. The complete assembly is immersed in a cooling bath. To ensure slow cooling the temperature of the cooling bath should be about 5°C below the freezing point of the test liquid. The test liquid should be stirred frequently during cooling since this minimises undercooling. Owing to the evolution of latent heat of fusion during the freezing process the temperature of the liquid will remain constant at the freezing point until the process has been completed. This constant temperature is noted.

To determine the freezing point of a pure liquid metal accurately a time–temperature cooling curve should be plotted from observed data. It is also necessary to plot cooling curves in order to determine liquidus and solidus temperatures for mixtures. The use of mercury in glass thermometers is not suitable in connection with liquid-metal determinations and the most convenient temperature-measuring device is a thermocouple connected to an accurate potentiometer. Although not as accurate, the principle of determining freezing points from a cooling curve may be demonstrated by connecting the thermocouple to a sensitive millivoltmeter instead of to a potentiometer. The reference or cold junction of the thermocouple must be kept at a constant temperature and this is best achieved by immersing it in melting ice in a vacuum flask. Low melting point metals, such as lead or tin, may be conveniently melted in a small crucible by means of a Bunsen flame.

The time–temperature curve for the freezing of a pure substance will appear as in figure 6.1a with a marked thermal arrest at the freezing point. If undercooling of the liquid occurs before the onset of freezing (dotted portion of curve) the latent heat emission when solidification commences will cause the temperature of the mass to rise to the true freezing temperature. In addition to pure metals, mixtures of eutectic composition and solid solutions of minimum or maximum melting point (figure 4.7b and c) will also freeze at constant temperature.

When time–temperature data are plotted for the cooling of a liquid solution the curve obtained will generally be similar to either figure 6.1b or figure 6.1c. A cooling curve similar to figure 6.1b would be shown by the following: alloys X and Y in figure 4.2, alloy (1) in figure 4.8 and alloy (1) in figure 4.9. In each of these cases solidification commences when the liquidus temperature appropriate to the particular

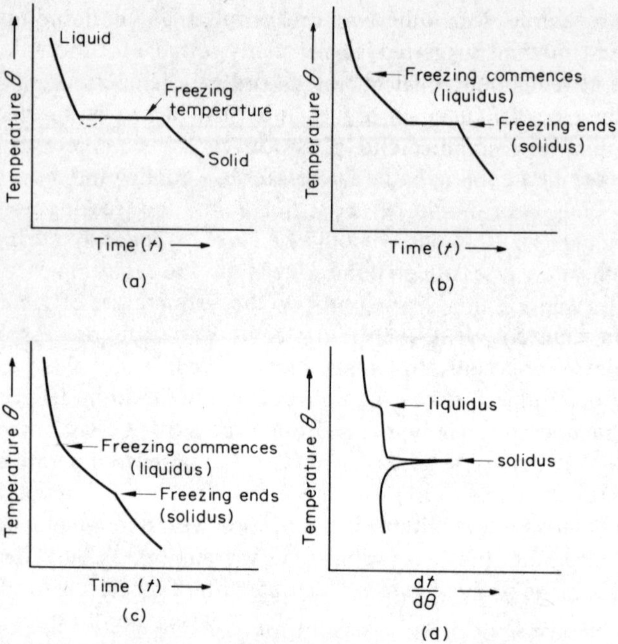

Figure 6.1. *Time–temperature cooling curves: (a) for a pure substance; (b) for a mixture with a eutectic; (c) for a solid solution; (d) inverse rate cooling curve for (b)*

alloy composition is reached. The first solid formed is of a different composition from the liquid solution. During the solidification process the liquid composition changes and the freezing temperature decreases. A discontinuity appears on the cooling curve at a temperature corresponding to the onset of freezing and the slope of the cooling curve is reduced. This is because latent heat emission during the freezing process retards the rate of temperature drop of the substance. In the examples quoted the final stages of solidification occur at a constant temperature, the eutectic temperature in the first two cases and the peritectic temperature in the last case. Both liquidus and solidus temperatures can be determined from this type of cooling curve.

The cooling curve shown in figure 6.1c is obtained when a solid solution type alloy freezes, that is any mixture composition in a phase system of the type shown in figure 4.7a or alloys (2) and (3) in figure 4.8. For this type of solution there is no horizontal portion in the cooling curve but only two gradient changes corresponding to the liquidus and solidus temperatures.

It is sometimes difficult to determine the exact temperatures at which a gradient change occurs in a time–temperature cooling curve but the liquidus and solidus points can be located with ease if the data are plotted in the form of an inverse rate-cooling curve, that is $dt/d\theta$ against temperature θ. This type of curve, corresponding to the cooling curve figure 6.1b is shown as figure 6.1d.

By making a complete series of representative alloys from pure substances and determining the cooling curves or inverse-rate cooling curves for each alloy, sufficient data may be obtained to plot the liquidus and solidus curves of a phase diagram. Lead and tin are very suitable materials for laboratory use owing to their low melting temperatures and ready availability.

6.3. Solvus Lines and Eutectoids

As stated earlier, there is an energy change, or heat of reaction associated with any phase reaction or transformation. This may involve a heat emission or heat absorption and applies to phase reactions involving the precipitation of a second phase from saturated solution or eutectoid transformations, as well as to phase reactions involving a chemical change.

It is possible to determine solvus and eutectoid phase-boundary lines for solids using a simple thermal-analysis technique similar to that described above for liquidus and solidus determinations. Heats of reaction are generally much smaller than those associated with fusion and a straight time–temperature cooling-curve plot will not give a clear indication of the phase transformation temperatures. Transformation temperatures should, on the other hand, be clearly defined on an inverse rate plot ($dt/d\theta - \theta$).

An interesting laboratory experiment to determine the critical temperatures of a steel can be based on this type of thermal analysis. A sample of a steel contained within a small electric furnace may be heated and cooled through the critical temperature range and time–temperature data recorded. The rate of heating and cooling should not be too rapid and rates in the region of 5–10°C per minute will give good results within a reasonable time period. Temperature is determined using a thermocouple in conjunction with a sensitive potentiometer with the elapsed time in seconds being noted at millivoltage steps equivalent to temperature intervals of about 2°C. When the results are plotted as inverse-rate cooling curves the A_1 and A_3 temperatures are readily discernible. It should also be possible to detect the A_2 point or Curie temperature. The values obtained during cooling will differ from those determined during heating owing to thermal hysteresis effects.

A more sensitive form of thermal analysis is differential thermal analysis in which the material under test is heated in close proximity to a reference test piece in which no phase reactions occur over the temperature range involved. If the sample and reference material are close together and located symmetrically with respect to the source of heat they should both be at the same temperature. During slow heating or slow cooling they should both be at the same temperature except when there is a phase reaction occurring within the test material. It is customary to use a thermocouple located within the furnace hot zone to record furnace temperature and to allow for control of heating and cooling rates, together with a differential thermocouple in contact with the sample under test and the reference material. With a differential thermocouple it is possible to detect very small differences in temperature.

With some of the highly sophisticated equipment available today differential thermal analysis and the related technique of differential scanning calorimetry it is possible not only to determine transition temperatures but also to obtain quantitative values for heats of reaction and specific heats.

6.4. Dilatometry

Within the general classification of thermal analysis is the technique of dilatometry. When a substance undergoes a phase change resulting in a change of structure, such a change is normally accompanied by a change in volume. When α iron changes on heating into γ iron the body-centred cubic crystal structure transforms into the more densely packed atomic arrangement known as face-centred cubic. There is a considerable decrease in the volume of the iron accompanying this structural change.

A dilatometer is an instrument which will allow either changes in volume or changes in length of a sample to be measured. A simple type of dilatometer, which may be used for the determination of the transition point of a substance that transforms at a temperature not greatly in excess of ambient temperature, consists of a glass bulb of about 5000 mm^3 capacity with a fine-bore stem. The substance under test is introduced into the bulb and is covered with an air-free liquid. The bulb is placed in a heating bath and the temperature of the bath is raised slowly. The position of the liquid in the dilatometer stem is recorded at successive temperature intervals. Normal thermal expansion is noted, except at the transition temperature, when there will be a sudden increase or decrease in volume, depending on whether the new phase is less dense or more dense than the original.

Another type of dilatometer, which is most suitable for operating at temperatures up to 1000°C, consists of an outer silica tube holding the specimen under test. An inner silica tube rests on the upper surface of the sample and movement of the inner tube following expansion and contraction of the specimen is mechanically amplified and converted into a dial-gauge reading. The dial-gauge sensitivity should be such that length changes of 0.001 mm may be detected. A suitable sample size is 50 mm x 10 mm diameter with an axial hole of 3 mm diameter extending into the sample for about 20 mm. A thermocouple is inserted through the inner silica tube into the sample and connected to either a sensitive millivoltmeter or a potentiometer. The changes in length for successive temperature intervals may be noted during a slow heating and slow cooling cycle.

This experimental method is very suitable for the determination of the critical temperatures of steels, but unlike the thermal-analysis method discussed in section 6.3 it will not indicate the Curie temperature since no structural change occurs at this point.

In some materials there is no abrupt volume change but a sharp change in the temperature coefficients of such properties as volume. In this case a change in the slope of the expansion curve will be noted at the transition temperature. This type of change occurs at the glass transition temperature, T_g, of polymer materials. At temperatures above T_g there is sufficient thermal energy to allow for some movement of segments of molecular chains and the material possesses rubbery characteristics. At temperatures below T_g, a well-defined temperature for most linear polymers, the material is hard, brittle and glassy. There is insufficient energy for molecular movement to occur, the only motion being a vibration of atoms about their equilibrium positions.

The value of T_g for nylon 6.6 is about 60°C and this may be determined experimentally using a quartz-tube dilatometer of the type described above, but using a heating bath rather than a furnace as the source of heat.

6.5. Microscopy

Heat treatment coupled with microscopy may be used to determine the positions of phase boundary lines in many solid systems. For example, to establish the position of the solvus line FC in a phase diagram of the type shown in figure 4.8 a series of alloys may be prepared from pure materials and subjected to various heat treatments, followed by microscopic examination of prepared surfaces. Consider one alloy composition, say composition (3) in figure 4.8. Small samples of this composition are heated at various temperatures below the solidus and allowed to remain at temperature for a sufficiently long time in order that equilibrium conditions are established. Each sample is rapidly quenched in water to retain the high-temperature condition at ordinary temperatures, prepared for examination and viewed under the microscope. For samples of alloy composition (3) quenched from temperatures above t_2 the structure should be single phased and consist entirely of crystals of α solid solution. The structure of samples quenched from temperatures below t_2 should contain both α and β phases. In this way the solvus line temperature for this alloy composition can be established. Where two phases appear the relative proportion of α to β may be estimated. This method of investigation is not suitable for all alloy systems and cannot be applied to those materials in which the metastable phase, produced by quenching, spontaneously transforms into the equilibrium state. The method can be used, however, in systems in which a new non-equilibrium phase, such as martensite in steels, is created during quenching. For example, when a plain carbon steel is quenched from a temperature above the A_3 value, austenite is not retained but transforms into martensite. When the steel is quenched from some temperature between the A_1 and A_3 values, that is an $\alpha + \gamma$ structure at high temperature, the austenite portion of the structure converts into martensite while the ferrite remains unaltered.

Quenching from a high temperature followed by examination was used very considerably in years gone by to determine phase boundaries. While it is still a useful training technique for student experimentation, the need for it in research largely disappeared with the advent of the hot-stage microscope, in which the microscope stage is enclosed within a vacuum furnace, enabling microscopic examination of materials at elevated temperatures to be undertaken. One can also observe structural changes taking place during heating or cooling.

6.6. Use of X-rays.

When a monochromatic source of X-radiation is directed at a crystalline material, Bragg reflections will occur from the various lattice planes. Each reflecting plane will deflect a portion of the X-ray beam in accordance with the relationship

$$n\lambda = 2d \sin \theta$$

where λ = wavelength of the incident radiation;
d = interplanar spacing;
θ = incident angle of radiation with plane;
n = a small integer.

Only a small quantity of material is required and this, in the form of a fine powder, is situated in the path of the radiation. One of the best arrangements for this work is the Debye–Scherrer type camera, in which the powder sample is positioned at the centre of a circular camera. A strip of unexposed X-ray film is placed around the circumference of the camera. After exposure and development a series of lines appears on the film, each line corresponding to a particular reflecting plane within the crystals. Each type of crystal lattice produces a characteristic diffraction pattern on the developed X-ray film. Measurements taken from the film can be converted into Bragg reflection angles and the respective interplanar spacings calculated. From these values, with a knowledge of the crystal type, the lattice parameters may also be calculated. When a solid solution is formed the structure of the solvent is not changed but the presence of solute atoms of a larger or smaller size than the atoms of solvent will cause an alteration in the lattice parameter value. Hence, from X-ray measurements, solid solutions may be detected and evaluated. Also, the presence of even a small amount of second phase will result in a diffraction pattern for this phase being superimposed on the diffraction pattern for the primary phase in an X-ray powder photograph.

X-ray diffraction can be a very useful tool for helping to establish boundaries in phase diagrams.

6.7. Other Methods

There are many properties of a substance that suffer a change when a phase reaction occurs. In addition to crystal type, thermal expansion and density, which have already been mentioned, there may also be changes in the values of electrial and thermal conductivity, optical properties such as colour and refractive index, elastic moduli and damping capacity. Measurements of some or all of these properties may be used in the compilation of data for phase-diagram construction. Also, by regular or continuous monitoring of one or more of these properties, much information has been obtained on changes from metastable toward equilibrium states.

Vibration testing has been used to a considerable extent in research into tranformations of metastable states. There is a natural frequency of vibration for any object. This natural frequency is dependent on the shape and dimensions of the object and also on the elastic constants and density of the material. For a bar of circular or square cross section, clamped centrally and stimulated into longitudinal vibration, the following relationship exists

$$f = \frac{n}{2l}\left(\frac{E}{\rho}\right)^{1/2}$$

where f is the frequency, n is an integer, l is the length of the bar, E is Young's modulus and ρ is the density. With modern electrical excitation methods and electronic counting, the frequency of vibration can be measured with extremely high accuracy enabling constants such as Young's modulus to be measured to an accuracy of ± 0.001 per cent. Extremely small changes in E, resulting from structural changes within a slowly transforming material, can be detected.

The damping capacity of a material can also be determined to a very high accuracy in modern vibration tests. The damping capacity ζ of a material is the ability of a material to attenuate vibrations and is given by

$$\zeta = \frac{1}{n}$$

where n is the number of vibrations of a test bar in free attenuation from an amplitude A to an amplitude A/e, that is 0.368 A (e is the base of natural logarithms). Damping capacity is an extremely structure-sensitive property for any material and measurements of changes in damping capacity with time have been used in connection with precipitation studies in aluminium alloys and other materials.

Revision Questions

1. State the Phase Rule and discuss its significance and application, making reference to both one-component and multi-component systems.

2. Distinguish clearly between a monotropic and enantiotropic substance. Give one example of each type of substance.

3. Sketch a *p–t* diagram for a pure substance showing no polymorphic modifications. Explain the significance of both the triple point and the critical point, and discuss the continuity of the liquid and gaseous states. How does the *p–t* phase diagram for a pure substance that sublimes differ from that for a substance that melts when heated?

4. The activation energies for many thermally activated processes are of the order of 50 MJ/kmol. Show that the rate of a process with this activation energy is approximately doubled for an increase in temperature of 10°C near room temperature.

5. State the Arrhenius rate law for a thermally activated process. The rate of crystal growth in a substance, measured at several temperatures, is given in the table.

Temperature (°C)	Rate (m/s)
220	6.03×10^{-12}
310	3.55×10^{-8}
410	3.39×10^{-5}

Show that this data is consistent with Arrhenius's law and determine the activation energy of the process.

(*Answer* 229×10^3 J/mol.)

6. What is meant by the term allotropy? Illustrate your answer with reference to carbon, iron and sulphur. Describe a method that may be used to determine the transition temperature for a substance.

7. State Le Chatelier's principle. An increase in pressure causes an increase in the melting point of sulphur but has the effect of reducing the freezing point of water. Why is this so?

8. What is meant by the terms 'component', 'phase' and 'variance'? State the phase rule and give an account of the phase relationships in a liquid system comprising two partially miscible liquids showing an upper consolute temperature.

9. State Raoult's law for ideal mixtures. Show, by means of sketches, how positive and negative deviations from Raoult's law affect p–c and t–c relationships for non-ideal liquid mixtures.

10. Compare and contrast steam distillation and fractional distillation as possible methods for the purification of organic liquids.

11. Two liquids, A and B, are miscible in all proportions. The total vapour pressure curve for mixtures of A and B possesses a minimum value at a composition of 60 per cent of B. Discuss the distillation of liquid mixtures within this system.

    Can fractional distillation be undertaken for good effect with a liquid mixture containing 50 per cent of B and which component, if either, may be separated by distillation?

12. Two totally immiscible liquids, water and iodobenzene, are steam distilled under a pressure of 98×10^3 N/m^2. From the vapour pressure data given calculate:

    (a) the distillation temperature of the mixture;
    (b) the composition of the distillate.

Temperature (°C)	Vapour pressures (N/m^2 x 10^3) Water	Iodobenzene
70	31.1	1.81
80	47.4	2.9
90	70.0	4.46
100	101.5	6.7

Assume molecular mass numbers: water = 18; iodobenzene = 204.
(*Answers* (a) 97.5°C; (b) 42.6 per cent iodobenzene.)

13. The water–ethylene glycol system shows a *t–c* phase diagram with the same general features as the lead–tin phase diagram. Discuss the technical significance of this and how this knowledge may be used for the solution of an engineering problem.

14. Two hypothetical metals A and B have melting points of 750°C and 900°C respectively. Draw four neat phase diagrams, clearly labelling all points and areas, based on the following conditions:

(a) the two metals are soluble in each other in all proportions in the solid state;
(b) the two metals are partially soluble in each other in the solid state;
(c) the two metals form an intermetallic compound with the formula AB_2;
(d) the two metals are totally insoluble in one another in the liquid state.

15. Cooling curves for several alloys of two metals A and B yield the following data

Composition (%B)	Liquidus temperature (°C)	Solidus temperature (°C)
10	380	255
30	255	235
40	255	235
60	400	235
80	510	235

From this data draw the phase diagram for the alloy system of A and B and clearly label all areas. Describe the manner of solidification and the equilibrium room temperature microstructures for the alloy containing 10 per cent of B and the alloy containing 30 per cent of B.

16. Two metals A and B of melting points 900°C and 700°C, respectively, are completely soluble in one another when liquid but only partially soluble in one another when solid.

Metal B is soluble in metal A to the extent of 25 per cent at 450°C and 10 per cent at 200°C. Metal A is soluble in metal B to the extent of 20 per cent at 450°C and 5 per cent at 200°C. At 450°C there is a eutectic containing 60 per cent of B. Draw, and label fully, the phase diagram for the alloy system A–B.

For an alloy containing 30 per cent of B:

(a) describe the manner in which it solidifies;
(b) state the phases present, and their relative proportions, assuming that full equilibrium has been attained at the following temperatures (i) just above 450°C, (ii) just below 450°C, (iii) 200°C.

(*Answers* (i) α to liquid = 6 to 1; (ii) α to β = 10 to 1, or α to eutectic = 6 to 1; (iii) α to β = 13 to 4.)

17. Describe with the aid of diagrams

(a) substitutional solid solutions;
(b) interstitial solid solutions

Discuss the factors that affect the mutual solubility of metallic elements, and the effects that solute atoms have on the properties of the alloy.

18. Two substances, with melting points of 600°C and 900°C respectively, are soluble in one another in all proportions in both the liquid and solid states. Draw the phase diagram for the binary system and describe the mode of solidification of any one mixture of the two substances.

Explain the term cored structure as applied to solid solutions, indicating how this type of structure is formed and how it may be eliminated.

19. How does the constitution of alloys affect their physical and mechanical properties? Illustrate your answer with diagrams covering the basic types of alloy systems (binary only).

20. What is meant by the term cryohydric point?

Sketch and label a binary *t–c* diagram involving water and a soluble salt. What is the significance of a discontinuity or discontinuities on a solubility curve for a salt in water?

21. Explain the features that are necessary in a binary phase diagram if there is to be a possibility of making alloy compositions that will respond to precipitation hardening.

Discuss the various stages of an age-hardening or a precipitation-hardening process.

22. Describe the treatments that would be given to an alloy composed of aluminium with 4.5 per cent of copper to put that alloy into the condition of maximum strength.

23. A plain carbon steel containing 0.6 per cent of carbon is heat treated as follows:

(a) heated to 800°C and quenched in cold water;
(b) heated to 800°C and slowly cooled in the furnace;
(c) heated to 800°C, quenched in water and tempered at 300°C;
(d) heated to 800°C, quenched in water and tempered at 600°C.

Describe with the aid of diagrams the structures obtained by these treatments and indicate, in a general way, the sort of physical properties you would expect the different materials to possess.

24. What is martensite and why does this phase not appear on the Fe-Fe_3C phase diagram? What are the general properties of martensite and how may these be improved?

25. Discuss briefly two experimental methods that can be used for the determination of the critical temperatures of a steel. What precautions should be taken to ensure a reasonable accuracy? What is the significance of the results obtained?

26. What are the major differences between a binary metallic system and a binary ceramic system?
Silica (SiO_2) melts at 1710°C. The SiO_2-Na_2O system contains a eutectic. What is the commercial significance of this?

Index